Computational Hydraulics

Computational Hydraulics

Numerical Methods and Modelling

Ioana Popescu

IWA PUBLISHING

Published by **IWA Publishing**
Alliance House
12 Caxton Street
London SW1H 0QS, UK
Telephone: +44 (0)20 7654 5500
Fax: +44 (0)20 7654 5555
Email: publications@iwap.co.uk
Web: www.iwapublishing.com

First published 2014

British Library Cataloguing in Publication Data
A CIP catalogue record for this book is available from the British Library

ISBN 9781780400440 (Paperback)
ISBN 9781780400457 (eBook)

Contents

About the author

Ioana Popescu is Associate Professor of Hydroinformatics at UNESCO-IHE Institute for Water Education in Delft, The Netherlands. Her teaching and research focuses on computational methods, aspects of flood modeling and vulnerability related to floods, lake and reservoir modeling, river systems and water supply systems modelling and optimisation. Dr. Ioana Popescu has been involved in several EU FP7 research projects (IceWater, lenvis, EnviroGRIDS, Floodsite) as well as other research projects related to development and application of modelling systems for water related areas. She is also the author of more than 30 peer-reviewed journal articles and 40 other publications, including invited book chapters and conference proceedings.

Preface

'No modern applied mathematician, physical scientist, or engineer can be properly trained without some understanding of numerical methods'

—E. Issacson and H. B. Keller

Computational hydraulics is one of the many fields of science for which computers opened a new way of working. The user of computer based models should be able to critically evaluate results given by different computational methods used in solving practical hydraulic problems. As such the book is intended to serve as a course for both undergraduate and graduate students in the fields of hydraulics and water resources.

The text of this book is based on the course of "Numerical methods for free surface flow", which evolved over the years at UNESCO-IHE Institute for Water Education. My students have been a source of motivation in trying to explain and clarify complex aspects of numerical methods, as well as free surface flows and hydrodynamics of large water bodies such as lakes and reservoirs.

There is a great deal of literature devoted to computational hydraulics in terms of numerical methods and fluid flow phenomena. However, it would be impossible to cover everything in one single book. Hence I have tried to cover only those methods that are most used in hydrodynamics. In Chapter 1, the book introduces the concept of modeling and the contribution of both numerical methods and numerical analysis to modeling. A description of the basic hydraulic principles, and the problems addressed by these principles in the aquatic environment is presented in Chapter 2, followed by the discretization principles of the fluid flow domain in Chapter 3. The finite difference method and finite volume method are presented in Chapters 4 and 5, respectively, with all the necessary steps for building and applying these numerical method approaches in hydraulics. Chapter 6 presents the properties of numerical methods that gives insight into how to avoid wrong results. The final two chapters show different example applications of computational hydraulics: river system modeling alongside hydrodynamic and water quality modeling of lakes and reservoirs. Worked-out examples that help in understanding the main concepts of computational hydraulics are developed for demonstration purposes at the end of Chapters 4, 5, 6 and 7.

Many concepts and results of the book have been discussed over the past years with Andreja Jonoski and Roland Price, whom I would like to thank for their inspiring talks. Their encouragement and research collaboration has been invaluable to me. I would like to thank my students Mario Castro Gama and Quan Pan, who provided me with some of the figures in Chapters 7 and 8. Lastly I am grateful and I would like to acknowledge my family for kindly helping and supporting me during the book writing process as well as drawing most of the figures.

Ioana Popescu
Delft, 2014

Water is the driving force of all nature.

-Leonardo da Vinci -

Chapter 1

Modelling theory

1.1 CONTEXT AND NATURE OF MODELLING

Models come in many different forms, some of which we do not usually refer to as models. Language, for example is a model, making associations between abstract concepts and labels (defined by words), using a set of rules for clustering words together (grammar), as such allowing to describe reality (i.e., build models of reality).

A model is a simplified, schematic representation of the real world. Models are meant to help engineers, scientists and decision-makers to determine what is happening in reality and to predict what may happen in the future. In particular, they are useful for the assessment of the impact of human activities on the environment or on artificial systems. Models are covering a large range of problems, hence it is difficult to find one that fits all problems.

A classical definition of the model is 'a simplification of reality over some time period or spatial extent, intended to promote understanding of the real system' (Bellinger, 2004) or 'A model is a simplification of reality that retains enough aspects of the original system to make if useful to the modeller' (Chapra & Canale, 2006; Eykhoff, 1974). In this context the system is defined to be a part of reality (isolated from the rest), which consists of entities that are in mutual relationships (processes) and have limited interactions with the reality outside of the system. Examples of systems are ecosystem or a hydrological system. A model is a physical or mathematical description of a physical system including the interaction with the outside surrounding environment, which can be used to simulate the effect of changes in the system itself or the effect of changes due to conditions imposed on the system.

Modelling implies the construction of a model, or working with an existing model, while simulation means to mimic the system on a computer. Building a small prototype of a system, in order to test theories and design techniques was one of the first uses of a model in civil engineering and it is referred to as *physical modelling*. As science and technology progressed, so did the knowledge about systems and properties of a system could be quantified by mathematical equations that were either empirical or derived from basic principles. The exact solution of such equations is referred to as *analytical model*. Mathematical representations in form of differential equations do not always have explicit analytical solutions and the use of approximate numerical approaches, to solve complex equations, is referred to as *numerical modelling*.

Modelling is as much an art as it is a science. The purpose to construct a model is to gain insight into the underlying causes or to provide accurate prediction of the system behaviour. Reliability on the model results depends on the expertise of the modeller about the system to be modelled. In general models are, as mentioned, mathematical representations, in forms of equations, of physical phenomena, embedding parameters and quantitative relationships between different variables and parameters. They should be seen as condensed versions of the knowledge of the modeller about the modelled system.

Approaches to construct a model are different from one problem to another as well as from one modeller to another. An engineer approach to problem solution is different from that of a mathematician. Mathematicians are interested in finding whether a solution to a differential equation exists, while an engineer assumes that the existence of a physical system is proof enough for the existence of a solution and focuses in finding the solution itself. The understanding of an engineering system is essentially gained by observation and experiment. After years of observation and measurements, engineers notice that certain aspects of their studies occur repeatedly. Such general behaviour can then be expressed as fundamental laws or models. These kinds of models embody the cumulative wisdom of past experience. As the purpose of modelling is to increase the understanding of a system behaviour, the validity of it does not reduce to its fit to observations and measurements, but also to model's ability to simulate and replicate situations or data beyond those originally described in the model (James & Burges 1982; Dooge, 1986). For example a descriptive model of the system is used to understand how the system could work; or to predict how an unexpected event could affect the system; or in some cases to try different control approaches in case a system is controlled by structures.

The developer of a model considers that advances in modelling can only be attained by forming close links between measurement, theory and modelling (Woods *et al.,* 1995). Models are constantly being improved and advanced, which implies that users of the models should always obtain the up-to-date version of a model to be applied for finding solutions to an engineering problem (Schulze, 1995).

Models in civil engineering refer to the modelling of the continuum, however the present book is concerned with the modelling and numerical methods applied to fluid flow problems. Computational fluid dynamics (CFD) is defined as the field of science that uses computers and numerical techniques to solve problems involving fluid flow. The term Computational Hydraulics was defined by Abbot and Minns (1998) to be 'the reformulation of traditional hydraulics to suit the possibilities and requirements of discrete, sequential and recursive processes of digital computation', and it is a sub-field of CFD.

In order to build a model apart from the mathematical formulation of the phenomena a lot of observed data is required to perform simulations. For the majority of hydraulic problems there is an imbalance between the lack of available data (little knowledge about the problem) and the availability of sophisticated methods that can be applied to simulate the system under study. Numerical methods and simulations fill the 'gap' between theory and experiment, providing qualitative and quantitative insight into many phenomena that are too complex to be dealt with by analytical methods and too expensive, time consuming and/or impossible to be performed experimentally. Hence it can be stated that simple models provide insight, complicated models provide many results for which insight need to be carefully analysed.

1.1.1 Classification of models

Classification of models is done in order to understand the type and level of mathematics involved in developing a model and it helps understanding how the simulation is done (Lane & Nichols, 1993). Moreover knowing the mathematics employed the user of a model can determine the nature of the output, as well as the amount of information required to build the model. Model classification helps users to select the appropriate model for a particular problem solving need.

Before any classification of models is done several formal mathematical notions and definitions are introduced:

- independent variables used in modelling are space and time. Time is usually defined over an interval, $t \in [t_o, T]$, and space x refers to the volume $\boldsymbol{V}$ that contains the system under study, $x \in \boldsymbol{V}$;
- dependent variable is the state variable, which takes values depending on parameters and independent variables. The state variable is a finite dimensional vector $\boldsymbol{u} = \boldsymbol{u}(x, t)$, which is n dimensional $\boldsymbol{u} = (u_1, u_2, \dots u_n)$, and describes sufficiently the evolution of the phenomena (real system);
- a mathematical model is the set of equations that defines the evolution of the state variable in space and time.

In order to properly represent the phenomena in nature, the real systems and the mathematical model need to be consistent. This means that the number of unknown dependent variables must be equal to the number of independent equations.

Purely from mathematical point of view models are *dynamic* or *static*, or they are *finite* and *continuous*. A mathematical model is *dynamic* if the state variable $\boldsymbol{u}$ is time dependent. If $\boldsymbol{u}$ does not depend on time then the mathematical model is *static*. A mathematical model is *finite* if the state variable does not depend on the space variables. Otherwise the mathematical model is *continuous*. Finite dynamic models are represented through ordinary differential equations; while continuous dynamic models are represented by partial differential equations. Static models, finite or continuous, are particular cases of dynamic models (the case when time derivative is zero).

From engineering point of view mathematical models for water related problems can be described as:

- Lumped conceptual models that are based on the concept of exchanges between global storage entities, which compose the system under study (i.e., one compartment for surface water, another one for aquifer storage, etc). These models satisfy principles such as the continuity principle, but do not embed a complete description of the driving forces, which govern the system to be modelled (e.g., rainfall-runoff model, reservoir model, etc).
- Physically-based, distributed models, that use a description of the physical phenomenon which govern the behaviour of water in the system under study. The principles that are applied are mass conservation and additional laws describing the driving forces, such as momentum equation. In case that these models refer to flow in saturated or unsaturated porous media, along with the continuity principle, the hypothesis of laminar flow is applied (i.e., shear stress proportional to the velocity, which determines the Darcy or Richards' equations for flow).
- Data driven models that seek for a correlation between input and output data, without trying to detail/analyse phenomena (e.g., linear regression, unit hydrograph). These models do need a lot of prior knowledge and data observations of the system under study.

Lumped conceptual and physically-based models use differential relationships, describing the changes in time of a set of variables. Most of the time, the formulations used are either Ordinary Differential Equations (ODEs) or Partial Differential Equations (PDEs). As these ODEs and PDEs are very often complex, it is not always possible to find analytical solutions to them.

Researchers (Singh, 1995; Tim, 1995), further classified models depending on the description of:

- Problem type and solution (Figure 1.1).
- Processes addressed (water quality, water allocation, reservoir operation, flood management, etc).

- Time and space problem dimension (lumped conceptual, one dimensional, two dimensional and three dimensional) (Figure 1.2);
- Method of solution (finite differences, finite elements, etc. See Figure 1.3.)

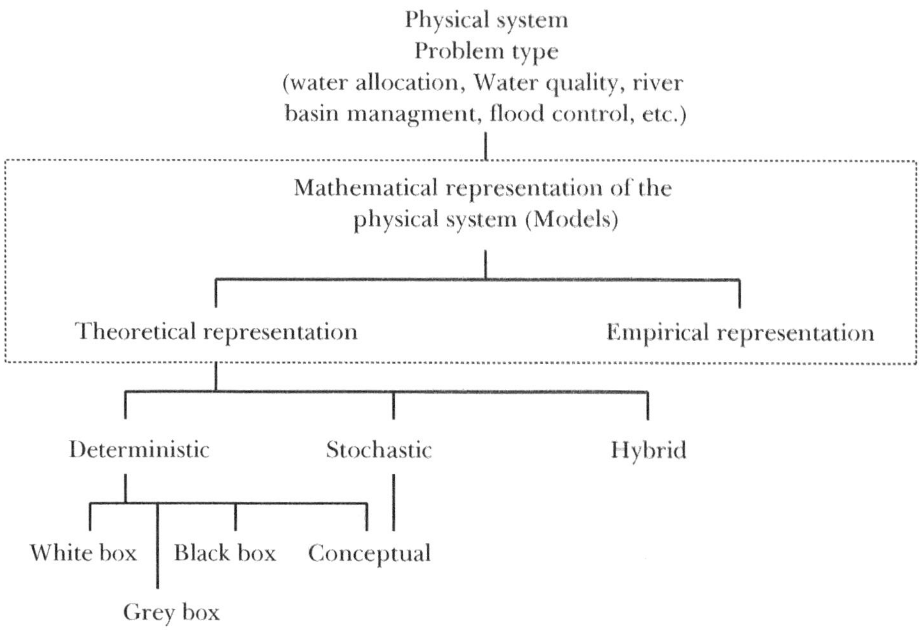

Figure 1.1 Classification of models based on model type.

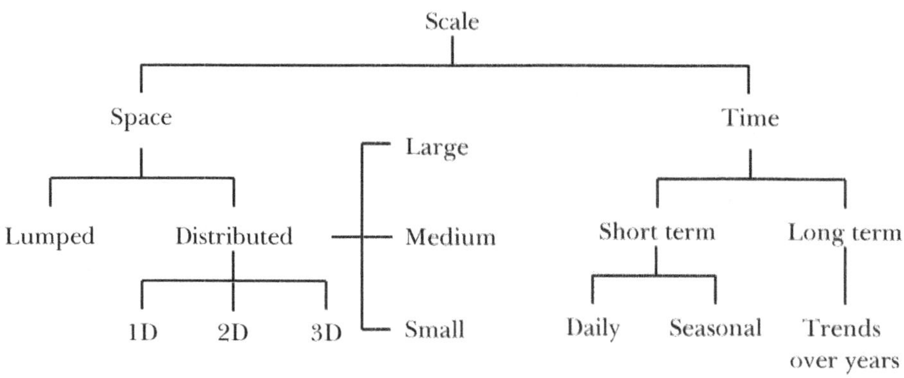

Figure 1.2 Classification of models based on scale.

Mathematical modelling problems are structurally classified into *black box* and *white box* models, depending on how much information is available about the system under study before the formulation of the mathematical model to be used for the problem solution. A black-box model is a system for which no a priori information is available and no understanding of the processes involved in the transformation is required. Only the input and the output have physical meaning. Stochastic models are black box models. A white-box model is a system where all necessary information is available and all processes are understandable and accounted for. Deterministic and physically based models fall into this category. The term *grey box* can also be used if not enough data is available and just partial understanding of processes is available. Lumped conceptual models are grey box models. In practice it is good to use as much a priori information as possible because it makes models behave correctly (accurately) (see Figure 1.4).

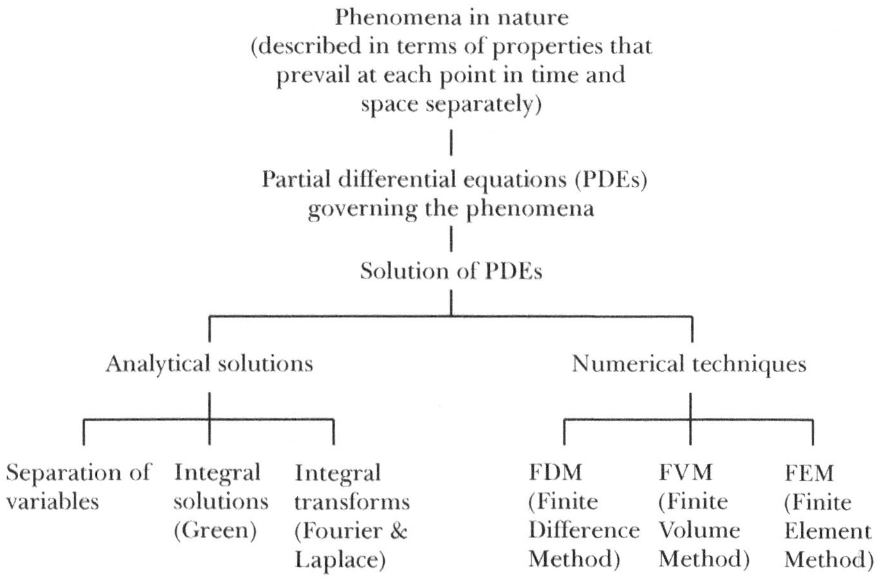

Figure 1.3 Classification of models based on method of solution.

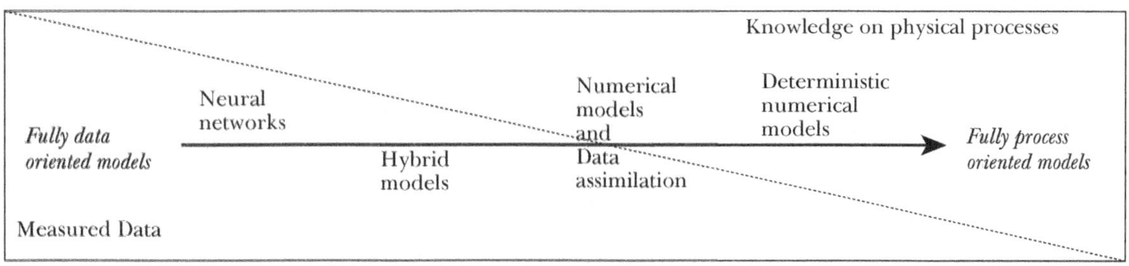

Figure 1.4 Models coverage based on data availability.

1.1.2 Computational Hydraulics

The field of Hydraulics can be divided into four major areas: Theoretical, Applied, Computational and Experimental. Theoretical area deals with fundamental laws and principles of mechanics studied for their intrinsic value. The Applied area transfers the theoretical knowledge to scientific and engineering applications, especially as regards to the construction of mathematical models of physical phenomena. Computational Hydraulics solves specific problems by combining mathematical models with numerical methods implemented on digital computers, a process called simulation. Experimental fluid mechanics puts physical laws, mathematical models and numerical simulations under the test of observation. Computational Hydraulics is interdisciplinary and has four major contributing disciplines (see Figure 1.5).

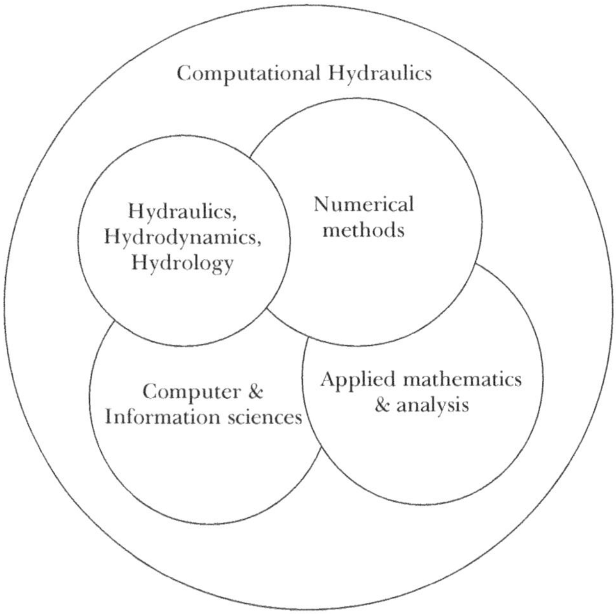

Figure 1.5 Computational Hydraulics field of study.

Engineering problems related to Hydraulics are of three kind; steady state, propagation problems and eigen-value problems. Steady state problems have as main characteristic the fact that the system does not change in time. Thus the equations representing the problem do not involve time as variable. Propagation problems are also called dynamic problems and the main characteristics of it is that the system changes with time. The variables and equilibrium relations depend in this case on time. The objective of the analysis is to calculate the state variables in time. In the case of eigen-values problems there is no unique solution to the system equilibrium equations. The analysis of the system has as the objective to determine the various possible solutions. Eigen-value problems arise in both steady-state and dynamic analysis.

1.2 CONCEPTUALIATION: BUILDING A MODEL

In general, mathematical modelling consists of several steps, as shown in Figure 1.6.

The starting point for any mathematical model is the Problem Description when the definition of what are the physical elements that are relevant in describing the phenomenon and the processes involved are defined. This stage of modelling requires input from experts in computational hydraulics and sometimes from other disciplines.

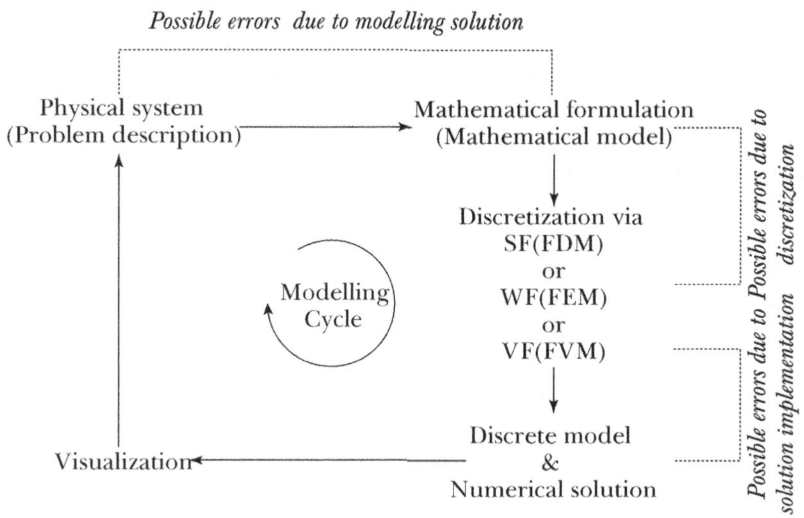

Figure 1.6 The cycle of mathematical modelling.

The next step is the Mathematical Formulation of the problem. After determining all relevant aspects, these are translated into mathematical relations that are in the form of a system of partial differential equations for various field quantities, which have to be solved over a given suitable domain (e.g., the whole volume of a lake, the whole river length). Boundary conditions are required over the domain of computation.

The third step is the one of translating the problem from the language of mathematics into a computable form. This is done by generating a grid of the computational domain on which the mathematical equations are formulated in a discrete form using different numerical approaches such as finite differences (FDM), finite volume (FVM), finite elements (FEM), boundary elements (BEM), and so on. The grid generation itself poses challenges for complex geometries applications. The discretization method is a matter of choice of the modeler and in majority of the cases depends on experience, but it should correspond to the grid used.

At the end of the discretization step a system of nonlinear or linear algebraic equations is obtained and appropriate solvers have to be used in order to find the solution. In case that nonlinear system of equations are obtained iterative methods based on successive linearization are used. This approach may lead to large systems of linear equations that are again solved using iterative methods. This step is known as Numerical Solution.

After obtaining a *solution*, visualization of results is needed. Post-processing is used to visualize the simulation results. This step enables interpretation of results, which is very important not only for the practitioner, but also for the developer of a model, because it helps to determine if there are errors in the computations and to see what is the quality of the model.

If results of simulation differ from the real situation, the model and/ or the numerical methods have to be refined and the cycle of mathematical modelling starts again.

1.3 MATHEMATICAL MODELLING IN PRACTICE

1.3.1 Selecting a proper model

Because of the development of numerical approaches many simulation techniques have been developped which lead to the proliferation of different models. As already mentioned models are intended for specific applications, however there are many models that are very general and have a wide area of applicability.

Naturally the question arise which model to choose from the wide variety of available models. The choice is difficult because there are models that are licence free as well as commercial ones.

Several authors (Woolhiser & Brakensiek, 1982; Sorooshian, 1991; Burnett & Watson, 1995) are defining practical criteria to be considered in selecting a model; the ability of the model to address the nature of the problem to be solved; availability of data for a particular problem; cost of the selected tool; possibility of applying the model for other similar problems.

The first criteria to consider is the ability of the model to address the problem that needs solution, by looking at the processes that are represented in the model, what are the equations representing these and what are the assumptions and limitations of the mathematical model. The availability of data is a very important issue, because even if the model is capable to describe processes that would be interesting for the modeller, if there is no data available to instantiate the model then a simpler model has to be considered.

Models range from very simple formula to complex ones that require implementation on a computer. there are advantages and disadvantages in using a complex or a simple model, but the most important is the objective for which the model is build. The choice to choose a complex model is an important decision in the modelling process, in the light data requirements, model applicability and computer power (Bergström, 1991). Complex models should be used for complex simulations, such as water quantity and quality for an entire catchment, however they should be used with care because they require a lot of input and there is a danger to introduce errors (Hughes & Beater, 1987). Complex models need long computational times and a thorough understanding of the model.

1.3.2 Testing a model

Any mathematical model is applied to solve engineering problems needs to be tested through validation, calibration and verification.

Schultze (1995) mentions that verification and validation notions are two terms used interchangeably by modellers, 'what is one modeller's verification may be another modeller's validation'. The use of the two terms might be misleading (Bredenhoeft & Koniko, 1993), hence the definition of the notions, according to the Oxford English Dictionary is given below in order to clarify terminology:

- Validate is defined as 'well founded and applicable, sound and to the point, against which no objection can be fairly brought' (i.e., authentic, true);
- Verify is defined as 'to test the accuracy, or establish the truth or correctness, of something by examination or by comparison with known data or some standard' (i.e., evidence, approval).

Any mathematical model selected to seek the solution for a given problem is first validated by using it to simulate small scale simple cases for which the results are known or can be easily obtained, analytically or by measurements. The validation ensures the applicability of the mathematical model. After selection of the model, the particular problem case is modeled by instantiating the model with data. The evaluation of the performance of it will take place in two steps; calibration and verification. During the calibration step all model parameters are estimated based on comparison of the simulation results with observed data. In the calibration phase the ability of the model to reproduce the response of the system is tested, as it is for example the simulation of a downstream hydrograph based on an inflow hydrograph upstream. The next step is verification when a simulation is performed, using parameters values obtained during calibration. The results of the verification step are compared with an observed data set, specifically reserved for this purpose.

In case of hydraulic and hydrological models, where time is involved, model calibration is usually done in three steps:

(1) *Selection of a simulation time period* from the available observation data set. Normally the available data set is split in two, one set for calibration and one for verification. Selection of the time length for calibration depends on the problem that needs to be addressed by the model. For example if flood is of interest the observed period should contain a flood event.
(2) *Preliminary calibration*, when based on the experience of the modeler a set of parameters are chosen and changed in such a way that the simulation results are as close as possible to the observed data.
(3) *Refined calibration*, when a thorough analysis of the output results is done and refinement of parameters values is carried out (either manually or by automatic procedures).

In order to determine the quality of the calibration a set of criteria are chosen, based on which the fit between simulated and observed data is evaluated. The selection of criteria depends on the objective of the modelling. In flood event modelling, for example several criteria such as time to peak, peak of the hydrograph, are used to assess the correspondence between observed and simulated flows.

Parameter calibration is not an easy task because some parameters may be influencing each other (especially in complex models) and this can generate sub-optimum determination of parameters. It is advisable that some of the parameters are obtained by direct field measurement, where possible, and for models for which they have physical meaning. Bathurst (1986) advises that a sensitivity of the fitting criterion to any changes in parameter values should be carried out before selecting any value for a parameter.

Immediately after calibration verification is carried out, by using the remaining data sets from the available observations. While verifying the outputs of a simulation the assumption is made that the governing equations and computer coding are valid, because the model passed the validation test. Verification therefore needs to be objective, subject to rigorous testing criteria, while taking into account the conditions for testing (e.g., years, inflow flood hydrograph, land use, etc).

With the development and application of models a special issue arises, the uncertainty of the model results. Uncertainty in modelling is of three types; structural, parametric or stochastic (Smith & Kuhn, 2002). Structural uncertainty comes from lack of knowledge about the phenomena that is modeled (i.e., lack of measurements, poor measured data, human errors, etc), while parametric uncertainty comes from using parameters with unsure values. Stochastic uncertainty shows the randomness of the phenomena and can not be reduced.

Though uncertainties are important in modelling, models can be evaluated based on clear defined criteria.

1.4 DEVELOPMENT AND APPLICATION OF MODELS

As a summary of the notions presented in this chapter a framework for development and use of models in engineering practice is presented bellow and schematized in Figure 1.7.

Several authors (Branson *et al.*, 1981; Shepherd & Geter, 1995) have outlined the steps that need to be taken in order to develop a proper modelling approach for solving waer related problems. These steps are:

(1) Define the objectives of the model, which requires that the modeler addresses the issues that are posed by the problem to be solved and determines what is the type of problem (water quantity, water quality, flood volume, etc);

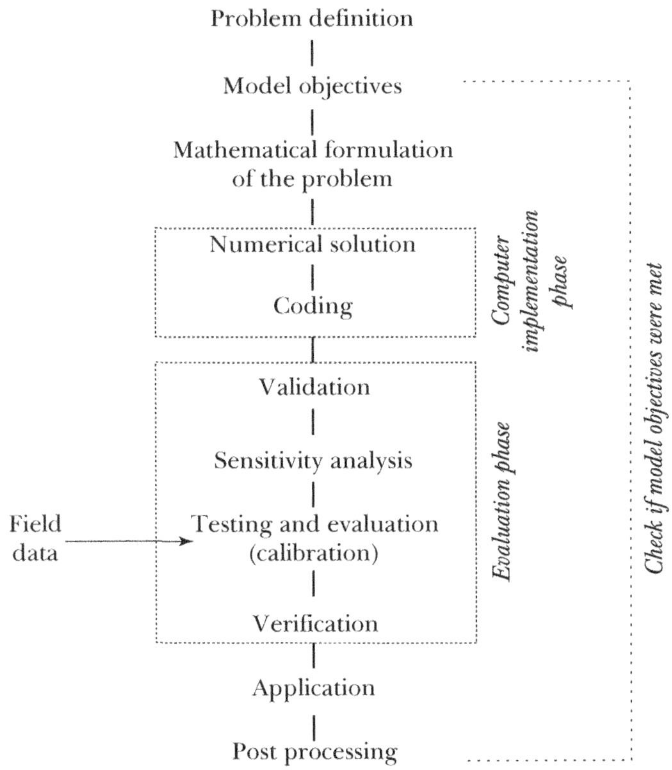

Figure 1.7 Model development and application steps.

(2) Determine if there is available a similar model that can be used. If there is no model available than a new model is defined and implemented on a computer;
(3) Define the mathematical representation of the phenomena that requires solution;
(4) Select the numerical approach that solves the mathematical representation of the phenomena;
(5) Coding and implementing the code on a computer;
(6) Validation of the mathematical model, by checking the model against simple examples. There is a need to check what are the limitations in applicability, and what are the required boundary and initial conditions;
(7) Sensitivity Analysis for the parameters of the model;
(8) Testing and Evaluation, when calibration and verification is performed;
(9) Application of the model for different problems;
(10) Presentation of results in graphical format, tabulated format or animated form to decision makers and stakeholders.

A good modelling practice does not mean just the development of a model and its implementation on a computer, but also continuous maintenance and refinement (improvement) without affecting its integrity.

Chapter 2
Modelling water related problems

2.1 BASIC CONSERVATION EQUATIONS

Computational fluid dynamics (CFD) is an important tool in engineering problem solving practice, covering a whole range of disciplines. As described in the previous chapter, Hydraulics is just one of the application fields of CFD, and it encompasses the study of free-surface and pressurized water flows. Traditionally Hydraulics is the scientific basis for engineering applications (Jain, 2001; Rousse, 1950). Hydraulics refer to water in both its states; still and flowing. The branch of Hydraulics concerned with the study of water flowing is referred as Hydrodynamics or Fluid Dynamics, when the studied fluid has different properties than water. The scope of the book is to address the equations representing the flow of water.

Conceptual models, based on empirical relations tested through observations (field measurements or laboratory experiments) are the first types of mathematical representations that were used in Hydraulics. Empirical relations were not the only representations used to describe behaviour of a water element; mathematical formulations in form of equations, were also used, however these equations could only give descriptions of the water behaviour and were representing relations between different variables and properties of the water. These types of equations were too complex to be solved analytically. The development of computers allowed for the formulation of solutions of particular situations, which brought the understanding of essential features of phenomena.

In this chapter the general formulation of the equations of water related problems are formulated. Their solution approach using numerical approximation is detailed in the following chapters of the book. The general classification of the equations, from mathematical point of view, is also given, because of the specific properties that each of the equations has.

In continuum mechanics the equations representing the general behaviour of a continuum are referred to as conservation laws, and they are based on Newton's laws. In Hydraulics the laws are conservation of mass, energy and momentum equation. These laws are not applicable only to fluids, they are also valid for solids, however they differ in behaviour, therefore due to the scope of the book, only the water related equations are addressed.

The laws of physics that are used for determining the general principle of conservation laws are Newton's laws of motion and the first and second laws of thermodynamics.

Newton's laws can be stated as:

- *First law*: a body is at rest or continues in a uniform state of motion, unless acted on by an applied force.
- *Second law*: the net force acting on a given mass is proportional to the time rate of change of the product of mass and velocity (also known as linear momentum);
- *Third law*: action and reaction are equal and opposite.

It is the second Newton's law that introduces the notion of change of state.

The laws of thermodynamics can be stated as:

- *First law*: in a system thermally isolated by impermeable walls the work done in taking one state A of a system into another state B is entirely determined by the terminal states A and B. The internal energy difference between state A and state B is defined as the mechanical work done in taking either state A into state B or state B into state A;
- *Second law*: there is a tendency on the part of nature to proceed towards a state of greater disorder.

Based on the above laws the general principle of conservation laws is that the rate of change of a quantity $\boldsymbol{u}$ within a volume $\boldsymbol{V}$ plus the flux of $\boldsymbol{u}$ through the boundary A (noted as $f(u)$) is the same as the rate of production of $\boldsymbol{u}$ denoted by $S(u, t)$ (see Figure 2.1):

$$\frac{\partial}{\partial t}\int_V \boldsymbol{u} \cdot dV + \int_A f(u) \cdot \boldsymbol{n} \cdot dA - \int_V S(\boldsymbol{u},t)dV = 0 \tag{2.1}$$

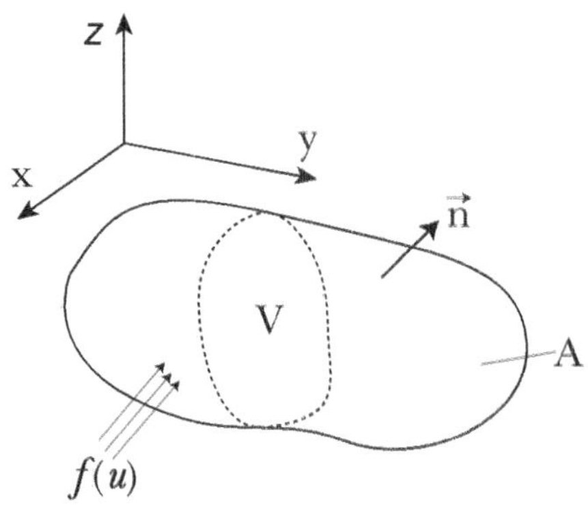

Figure 2.1 Control volume of quantity $\boldsymbol{u}$ in a (x, y, z) referential system.

Equation (2.1) is referred to as the *integral form* of the conservation law, and can be further detailed for mass, energy and momentum.

2.1.1 Conservation of mass

Conservation of mass states that for any control volume, during a small time interval Δt the mass entering the volume minus the mass leaving the volume equals the change of mass inside the control volume. In case of continuation of mass equation (2.1) for mass m of density ρ, and advection velocity $\vec{u}$ over the control volume V gives:

$$\frac{\partial}{\partial t}\int_V \rho \cdot dV + \int_A (\rho \cdot u) \cdot \boldsymbol{n} \cdot dA = 0 \tag{2.2}$$

where:

$$f(u) = \rho \cdot u \quad \text{and} \quad u \cdot \boldsymbol{n} = \vec{u} \tag{2.3}$$

In case of conservation of mass the term S, is the source term, and since there is no mass production, S is zero. Integration of equation (2.2) yields:

$$\frac{\partial \rho}{\partial t} + \nabla(\vec{u}\rho) = 0 \tag{2.4}$$

with $\nabla = \left(\frac{\partial}{\partial x}, \frac{\partial}{\partial y}, \frac{\partial}{\partial z}\right)$, the so-called *nabla* operator.

Equation (2.4) can be further detailed if the product rule for vectors is applied. It yields:

$$\frac{\partial \rho}{\partial t} + (\vec{u} \cdot \nabla)\rho + \rho\nabla \cdot \vec{u} = 0 \tag{2.5}$$

which can be re-arranged as:

$$\frac{D\rho}{Dt} + \rho\nabla \cdot \vec{u} = 0 \tag{2.6}$$

where D/Dt is the total derivative with respect to time:

$$\frac{D}{Dt} = \frac{\partial}{\partial t} + \frac{dx}{dt}\frac{\partial}{\partial x} + \frac{dy}{dt}\frac{\partial}{\partial y} + \frac{dz}{dt}\frac{\partial}{\partial z} \tag{2.7}$$

In general water is considered an incompressible fluid (i.e., $\rho = \text{const.}$), hence the continuity equation simplifies to:

$$\nabla \cdot \vec{u} = 0 \tag{2.8}$$

Equation (2.6) is the continuity of mass, while equation (2.8) can be regarded as the equation of continuity of volume.

2.1.2 Conservation of momentum

Conservation of momentum is demonstrated in x direction only. Conservation in y and z dimensions is similar.

In case of momentum in x direction, the conserved quantity is ρu_x. In equation (2.1) momentum flux in x direction is $\rho u_x \vec{u}$ and S are the body and boundary forces acting on the control volume. Body force is the gravity and boundary forces are pressure, shear and surface forces.

Momentum equation in x direction, according to equation (2.1), is:

$$\frac{\partial(\rho u_x)}{\partial t} + \nabla \cdot (\rho u_x \vec{u}) = \underbrace{S_{P,x}}_{Pressure\,force} + \underbrace{S_{F,x}}_{Friction\,Force} + \underbrace{S_{G,x}}_{Gravity\,Force} \tag{2.9}$$

If gravity and pressure forces are detailed, yields:

$$\rho \frac{Du}{Dt} = -\frac{\partial p}{\partial x} + \rho g_x + S_{F,x} \tag{2.10}$$

In three dimensions equation (2.10) becomes:

$$\rho \frac{D\vec{u}}{Dt} = -\nabla p + \rho \vec{g} + \vec{S}_F \tag{2.11}$$

where $\vec{u} = (u_x, u_y, u_z)$ and $\vec{g} = (g_x, g_y, g_z)$.

Using the total derivative definition, equation (2.11) becomes:

$$\frac{\partial(\rho \vec{u})}{\partial t} + \nabla(\rho \vec{u} \times \vec{u}) = -\nabla p + \rho \vec{g} + \vec{S}_F \tag{2.12}$$

In equation (2.12) the term $\vec{u} \times \vec{u}$ is the cross product, introduced, in calculus, by definition, as:

$$\vec{u} \times \vec{u} = \nabla \begin{bmatrix} u_x u_x & u_x u_y & u_x u_z \\ u_y u_x & u_y u_y & u_y u_z \\ u_z u_x & u_z u_y & u_z u_z \end{bmatrix} \tag{2.13}$$

Hence,

$$\begin{aligned} \nabla(\rho \vec{u} \times \vec{u}) &= \nabla \begin{bmatrix} \rho u_x u_x & \rho u_x u_y & \rho u_x u_z \\ \rho u_y u_x & \rho u_y u_y & \rho u_y u_z \\ \rho u_z u_x & \rho u_z u_y & \rho u_z u_z \end{bmatrix} \\ &= \begin{bmatrix} \frac{\partial}{\partial x}(\rho u_x u_x) + \frac{\partial}{\partial y}(\rho u_x u_y) + \frac{\partial}{\partial z}(\rho u_x u_z) \\ \frac{\partial}{\partial x}(\rho u_y u_x) + \frac{\partial}{\partial y}(\rho u_y u_y) + \frac{\partial}{\partial z}(u \rho_y u_z) \\ \frac{\partial}{\partial x}(\rho u_z u_x) + \frac{\partial}{\partial y}(\rho u_z u_y) + \frac{\partial}{\partial z}(\rho u_z u_z) \end{bmatrix} \\ &= (\nabla \rho \cdot \vec{u})\vec{u} + \rho \vec{u}(\nabla \cdot \vec{u}) + \rho(\vec{u} \cdot \nabla)\vec{u} \end{aligned} \tag{2.14}$$

Left side of equation (2.12) becomes:

$$\frac{\partial(\rho\vec{u})}{\partial t} + \nabla(\rho\vec{u} \times \vec{u}) = \rho\frac{\partial\vec{u}}{\partial t} + \underbrace{\vec{u}\frac{\partial\rho}{\partial t} + \vec{u}(\nabla\rho \cdot \vec{u}) + \rho\vec{u}(\nabla \cdot \vec{u})}_{\vec{u}\left(\frac{\partial\rho}{\partial t}+\nabla\cdot(\rho\vec{u})\right)=0-continuity\,equation} + \rho(\vec{u} \cdot \nabla)\vec{u} \tag{2.15}$$

Based on (2.15) equation (2.12) becomes:

$$\rho\frac{\partial\vec{u}}{\partial t} + \rho(\vec{u} \cdot \nabla)\vec{u} = -\nabla p + \rho\vec{g} + \vec{S}_F \tag{2.16}$$

Equation (2.16) represents the conservation of momentum in a three dimensional (3D) space.

2.1.3 Conservation of energy

For phenomena in which temperature varies, in addition to momentum equation, conservation of energy must be considered. The energy equation accounts for both kinetic and potential energy. Hence energy is expressed by the term $\rho\left(e + \frac{\vec{u}^2}{2}\right)$, and the equation for conservation of energy is:

$$\frac{\partial}{\partial t}\left[\rho\left(e + \frac{\vec{u}^2}{2}\right)\right] + \nabla \cdot \left[\rho\left(e + \frac{\vec{u}^2}{2}\right)\vec{u}\right] = \nabla \cdot (k\nabla T) - \nabla \cdot (\vec{u}p) + \nabla \cdot (\vec{u}\tau) + \rho\vec{g}\vec{u} \tag{2.17}$$

where e is the potential energy; p is pressure; τ is stress tensor; k coefficient; and T is temperature.

The three presented conservation laws form the so-called ***Navier-Stokes equations***, which are space and time-dependent, that is, there are four independent variables. In a three dimensional domain there are six dependent variables and only a set of five equations available; a continuity equation for conservation of mass, three equations for conservation of momentum and one equation for conservation of energy. The six dependent variables are pressure, density, temperature, and three components of the velocity vector ***u***. The six equation that is used is the law of thermodynamics relating pressure and temperature:

$$-p \cdot V = nRT \tag{2.18}$$

where p-pressure; V-control volume; T temperature; and n, R coefficients.

A system of six equations with six unknowns is formed, which can be solved. Usually, the Navier-Stokes equations are too complicated to be solved, and simpler forms are used. These are discussed further in this chapter.

2.2 MATHEMATICAL CLASSIFICATION OF FLOW EQUATIONS

One of the conclusions of the previous paragraph is that the physical processes related to water problems are represented mathematically by differential equations (commonly abbreviated to ODEs or PDEs, for Ordinary Differential Equations and Partial Differential Equations, respectively). A differential equation is a relationship among various distinct derivatives of a function of one or more variables. Examples of such processes are propagation of waves in a river channel, pollutant transport, reservoir operations, and so on. In describing these phenomena rate of change of a variable is evaluated, which results in the representation of the physical processes in terms of differential equations (DE). Differential equations are models of the phenomena and require integration to be solved. This is not always possible, hence they are solved numerically.

As opposed to an algebraic equation, where the unknown is a number, in a differential equation, the unknown is a function that expresses the behaviour of the variable as a function of time and space. A differential equation has order, which is given by the order of the derivative of the unknown function involved in the equation. If the values of the unknown function and its derivatives at some point are known the solution to DE is called an *initial value problem* (*IVP*). If no initial conditions are given, the description of all solutions to the DE is called the *general solution*.

DE are classified taking into account several criteria, as follows:

- dimension of the unknown function:
  - ordinary differential equation (ODE), when the unknown is a function of one independent variable only, for example, $u(t)$;
  - partial differential equation (PDE), when the unknown is a function of multiple independent variables, for example, $u(t, x, y)$;
- number of equations:
  - single differential equation; or
  - system of differential equations (coupled);
- order:
  - the highest order of the derivative that appears into equation, that is, an equation is of n-order if the DE has a n-th derivative, and not higher;
- linearity:
  - linear, when all terms are linear in unknown function and its derivatives;
  - non-linear, when the linearity does not hold.

In water related problems, most frequently the 2nd order, linear PDEs with two independent variables (homogeneous or nonhomogeneous) are used. In such a case, if the two independent variables are assumed to be x and t, the PDE is written as:

$$A(x,t)\frac{\partial^2 u}{\partial x^2} + B(x,t)\frac{\partial^2 u}{\partial x \partial t} + C(x,t)\frac{\partial^2 u}{\partial t^2} = D(x,t)\frac{\partial u}{\partial x} + E(x,t)\frac{\partial u}{\partial t} + F(x,t)u(x,t) + G(x,t) \tag{2.19}$$

where the coefficients $A(x, t) \ldots G(x, t)$ may be functions of x, t, or both, or they may be a constant. The function $u(x, t)$ is the solution of the equation. The second-order PDE (2.19) has a discriminant defined as:

$$\Delta = [B(x,t)]^2 - 4 \cdot A(x,t) \cdot C(x,t) \tag{2.20}$$

Depending on the sign of the discriminant Δ (as defined by equation (2.20)), equation (2.19) is classified as:

- elliptic PDEs, if $\Delta < 0$, which are PDEs with smooth solutions, easy to solve;
- hyperbolic PDEs, if $\Delta > 0$, which have solutions that may have discontinuities, usually difficult to solve computationally; and
- parabolic PDEs, if $\Delta = 0$, which may have features of hyperbolic and elliptic PDEs combined.

The terminology for this classification is coined from the geometry of conic sections that satisfies a similar type of equation like (2.19), just that they are expressed in coordinates x and y.

Each type of equation describes a certain type of physical phenomena; elliptic PDEs describe processes that are in steady state, time-independent (e.g., steady-state aquifer flow, backwater curves); hyperbolic PDEs describe time-dependent, conservative physical processes, (e.g., advection/convection), that are in unsteady state and not evolving towards steady state; and parabolic PDEs describe time-dependent, dissipative physical processes (e.g., such as diffusion, transient aquifer flow) that are evolving towards a steady state. These three types of equations are very important for flow problems and the solution for each one of them is further presented.

2.2.1 Solutions of ODE

An Ordinary Differential Equation (ODE) is a differential equation where the solution depends on one independent variable only. The general form of an ODE is:

$$\frac{du(t)}{dt} = f(u,t) \tag{2.21}$$

where $u(t)$ is an unknown function of the variable t (time) and f is an arbitrary function in $u(t)$ and t.

Ordinary differential equations may have analytical solutions that are easy to obtain if the function $f(u, t)$ is not too complicated. However even if the analytical solution exist the implementation of the solution into a computer programme may be difficult.

If along with equation (2.21) an initial value of the function $u(x, t)$ is given then the problem is called initial value problem (*IVP*) and can be solved easily using numerical approximations, as it is detailed further in Chapters 4 and 5 of the book.

2.2.2 Solutions of PDE

In order to have a solution for a PDE, the problem must be well posed mathematically (Garabedian, 1966; Cunge & Verwey, 1986; Ligget & Cunge, 1975). This implies that solution fulfils three conditions; it exists; it is unique; and depends on auxiliary conditions such as boundary conditions. Fletcher (1998) states that uniqueness of solution is not a problem in general, and if it does not exist is mainly due to failure to fulfil auxiliary conditions. Moreover from mathematical point of view for some particular equations there are isolated points in the definition domain for which the solution may not be unique. There are physical phenomena for which multiple solutions exists, due to the physical phenomena, not due to the mathematical position of the problem. Such an example is the case of transition from laminar to turbulent flow.

Any PDE can be solved if boundary conditions and initial conditions on the computational domain are known. Typical representations of the computational domains and required boundary conditions for the three defined types of second order PDEs are represented in Figure 2.2.

The definition of boundary conditions constitutes the start for determining the solution of any problem inside the computational domain. There are three ways of defining boundary conditions; Dirichlet type of conditions, when the values of the unknown function $\boldsymbol{u}$ are known at the border of the computational volume (on A, see Figure 2.1); Neumann type of conditions, when derivatives of the unknown function is known at the border of the computational domain; and Robin or mixed type of conditions, when a combination of Dirichlet and Neumann conditions are applied at the domain boundary.

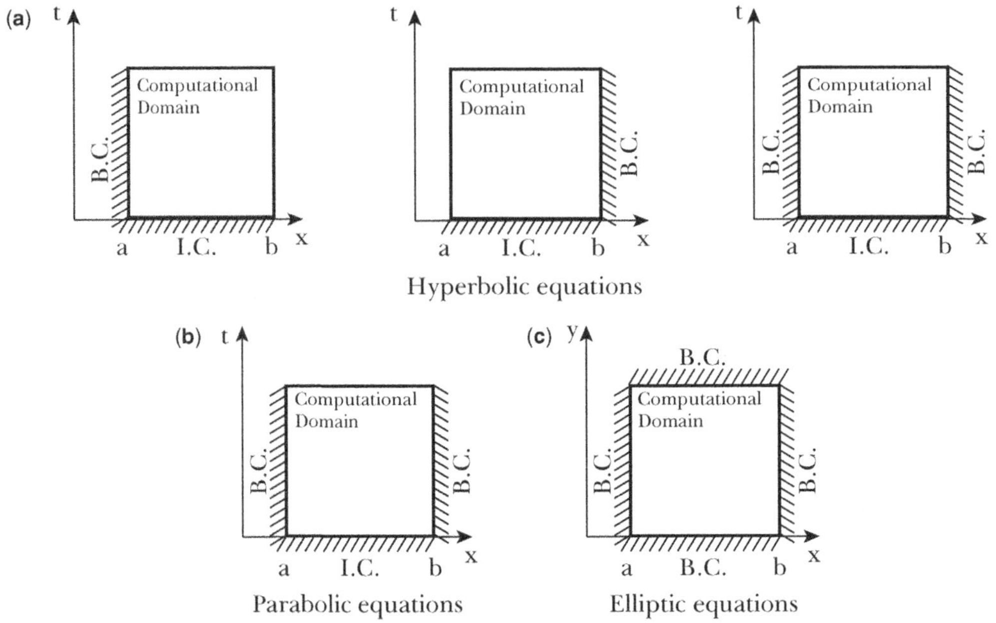

Figure 2.2 Computational domains, boundary and initial conditions for PDEs. (B.C. Boundary conditions; I.C. Initial conditions; (a) Hyperbolic type of equations; (b) Parabolic type of equations; (c) Elliptic type of equations).

An important issue in finding the solution of a PDE are the errors. Hyperbolic types of problems are the most exposed to errors because in their case conditions at the boundary of the computational domain are the ones introducing the errors that will propagate inside the computational domain.

2.2.2.1 Boundary conditions for fluid flows PDEs

Majority of the flow phenomena in hydraulics (hydrodynamics) are governed by a system of two differential equations (as it is for example the case of free surface flows). A system of two first order partial differential equations, having two independent variables (ξ, η) and two unknown functions (u and v), is written as:

$$a_1 \frac{\partial u}{\partial \xi} + b_1 \frac{\partial u}{\partial \eta} + c_1 \frac{\partial v}{\partial \xi} + d_1 \frac{\partial v}{\partial \eta} = f_1 \tag{2.22a}$$

$$a_2 \frac{\partial u}{\partial \xi} + b_2 \frac{\partial u}{\partial \eta} + c_2 \frac{\partial v}{\partial \xi} + d_2 \frac{\partial v}{\partial \eta} = f_2 \tag{2.22b}$$

where, ξ and η are the independent variables. For water related problems ξ corresponds to the x-coordinate and η to time t. The unknown functions are variables that correspond to flow quantities, such as flow velocity or water depth h.

Because both u and v are functions of (ξ, η), their total derivatives in the (ξ, η) plane are:

$$du = \frac{\partial u}{\partial \xi} d\xi + \frac{\partial u}{\partial \eta} d\eta \tag{2.23a}$$

$$dv = \frac{\partial v}{\partial \xi} d\xi + \frac{\partial v}{\partial \eta} d\eta \tag{2.23b}$$

These derivatives can have different values in different regions of the (ξ, η) plane, or may not exist in some regions. The curves that are splitting the lines into different regions are called characteristic curves and are determined from the condition that the determinant of the system formed from the four equations (2.22*a*, *b*) and (2.2.3*a*, *b*) is not zero, that is, the system of equations has a unique solution. The formed system of equations is written in matrix format as:

$$\begin{bmatrix} a_1 & b_1 & c_1 & d_1 \\ a_2 & b_2 & c_2 & d_2 \\ d\xi & d\eta & 0 & 0 \\ 0 & 0 & d\xi & d\eta \end{bmatrix} \begin{bmatrix} \partial u/\partial \xi \\ \partial u/\partial \eta \\ \partial v/\partial \xi \\ \partial v/\partial \eta \end{bmatrix} = \begin{bmatrix} f_1 \\ f_2 \\ du \\ dv \end{bmatrix} \tag{2.24}$$

The unknowns of the system (2.24) are the derivatives of the unknown functions u and v. These are not determined if the determinant of the system (2.24) is zero, The value zero of the determinant defined by (2.24) determines the regions in space beyond which there is no solution for the given equation, because the derivatives do not exists. The condition is written as:

$$\begin{bmatrix} a_1 & b_1 & c_1 & d_1 \\ a_2 & b_2 & c_2 & d_2 \\ d\xi & d\eta & 0 & 0 \\ 0 & 0 & d\xi & d\eta \end{bmatrix} = 0 \tag{2.25}$$

Calculating the determinant gives:

$$(a_1c_2 - a_2c_1) \cdot \left(\frac{d\eta}{d\xi}\right)^2 - (a_1d_2 - a_2d_1 + b_1c_2 - b_2c_1) \cdot \frac{d\eta}{d\xi} + (b_1d_2 - b_2d_1) = 0 \tag{2.26}$$

Rearranging yields:

$$a\left(\frac{d\eta}{d\xi}\right)^2 + b\frac{d\eta}{d\xi} + c = 0 \tag{2.27}$$

with coefficients:

$$a = (a_1c_2 - a_2c_1); \quad b = (-a_1d_2 + a_2d_1 - b_1c_2 + b_2c_1); \quad c = (b_1d_2 - b_2d_1) \tag{2.28}$$

Equation (2.27) has two solutions, real or imaginary, depending on the value of the discriminant. The two solutions of the equation (2.27) are the known solutions of the quadratic equation:

$$\frac{d\eta}{d\xi} = \frac{-b \pm \sqrt{b^2 - 4ac}}{2a} \tag{2.29}$$

Equation (2.29) represents curves in the (ξ, η) plane, called *characteristic curves* (C1 and C2), and determines the number of required boundary conditions in order to have a unique solution. The discriminant is similar to the one defining the types of PDEs, hence depending on the value of the discriminant the problem can be of 3 types: hyperbolic – with two real characteristics; parabolic – with one characteristic; or elliptic – with two imaginary characteristics (see Figure 2.3).

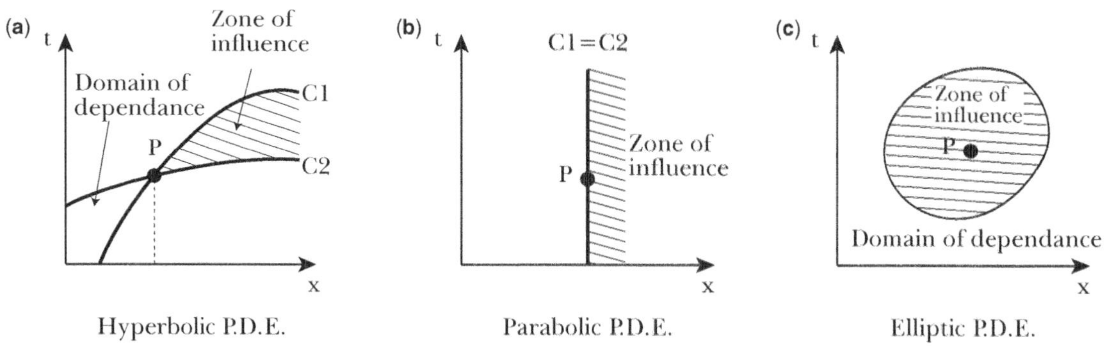

Figure 2.3 Domain of solution for a system of two first order differential equations.

If the analysis is done from the point of view of the solution of flow phenomena it can be stated, as already mentioned, that hyperbolic behaviour means that a system develop from the state at point *P* of the computational domain towards a state in between the characteristics C1 and C2 (shock waves). In case of parabolic type of problems it can be seen that from a state *P* in the computational domain the solution can evolve only in one part of the computational domain. Elliptic type of problems can develop from any state *P* to any state in the computational domain.

2.2.2.2 Hyperbolic equation examples

The typical example of a hyperbolic equation is the wave equation:

$$\frac{\partial^2 u}{\partial t^2} - a^2 \frac{\partial^2 u}{\partial x^2} = 0 \tag{2.30}$$

where u is the unknown function and a is the wave speed, considered a constant in this example.

Given the initial and boundary conditions, over the domain (0, L) (see Figure 2.4a):

$$\begin{cases} u(x,0) = \sin \pi x \\ \left.\dfrac{\partial u}{\partial t}\right|_{(x,0)} = 0 \\ u(0,t) = u(L,t) = 0 \end{cases} \tag{2.31}$$

an exact solution can be determined as:

$$u(x,t) = a \cdot \sin \pi x \cdot \cos \pi t \tag{2.32}$$

Representation in the (x, t) plane, of the solution given by (2.32), is shown in Figure 2.4b.

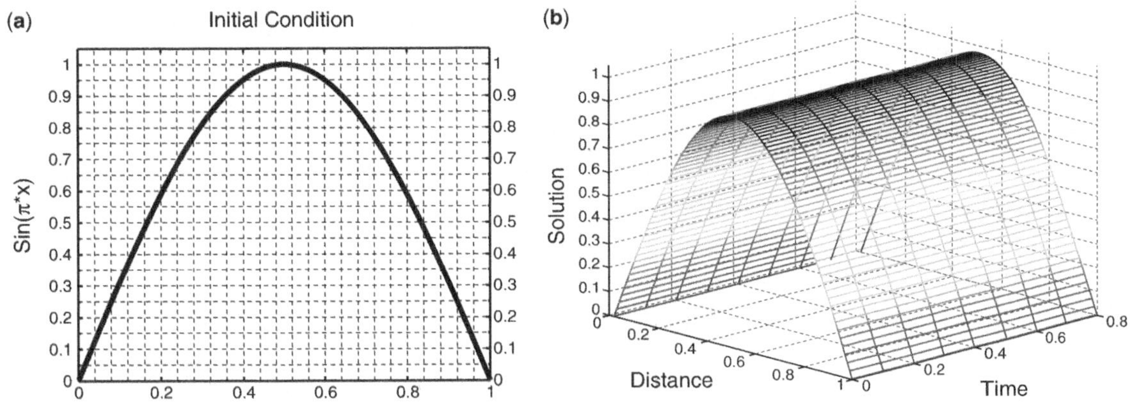

Figure 2.4 (a) Initial condition and (b) solution of the wave equation.

In general, for different types of initial conditions the solution of the equation (2.30) is not easy to be determined. The method used to solve the wave equation is usually the method of characteristics (Abbott & Basco, 1982). According to equation (2.27) there are two real characteristics available for equation (2.30):

$$\begin{cases} \dfrac{dx}{dt} = a \\ \dfrac{dx}{dt} = -a \end{cases} \tag{2.33}$$

The set of equations (2.33) are two straight lines (because a is a constant), which show that a disturbance in solution of u, taking place at point P influences the solution domain after P (CPD in Figure 2.5(b)). Solution in point P is influenced by disturbances in the domain before point P (APB). Depending on the position of P in the computational domain initial conditions may be sufficient to determine the solution at point P, uniquely. However there are points in the computational domain for which solution can be determined uniquely only if boundary conditions are given (see Figure 2.5).

Another example of hyperbolic equation, which is very common for many phenomena in aquatic environment, is the advection equation:

$$\frac{\partial u}{\partial t} + a\frac{\partial u}{\partial x} = 0 \tag{2.34}$$

In case of equation (2.34), just one characteristic line passes through every point of the space (see Figure 2.6). Equations (2.30) and (2.34), presented herein, are equation showing propagation phenomena. No dissipation effect is present, just convection.

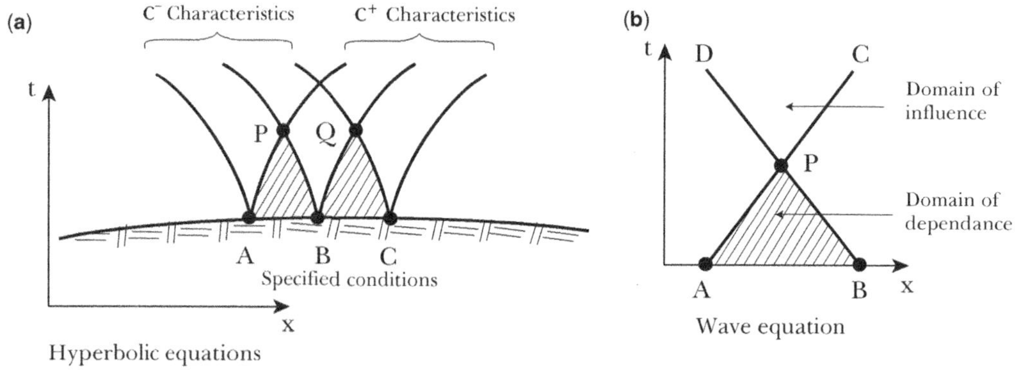

Figure 2.5 Characteristic curves for hyperbolic PDEs.

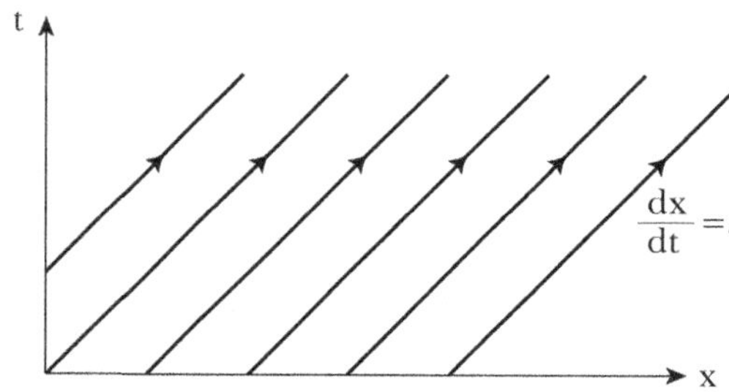

Figure 2.6 Characteristic lines for the advection equation.

2.2.2.3 Parabolic equation example

Phenomena in nature for which dissipation mechanism appears along with propagation are of parabolic type. The diffusion of a pollutant in a water body is such an example. The diffusion equation in one dimension of space is:

$$\frac{\partial u}{\partial t} = D\frac{\partial^2 u}{\partial x^2} \tag{2.35}$$

where D is the diffusion coefficient. Equation (2.35) is defined over the domain $(0, L)$ (see Figure 2.7).

Given the same initial and boundary conditions as in the case of wave equation (2.30), the analytical solution is:

$$u(x,t) = \sin \pi x \cdot \exp(-\pi^2 t) \tag{2.36}$$

The exponential part of the solution given by equation (2.36) represents the time decay.

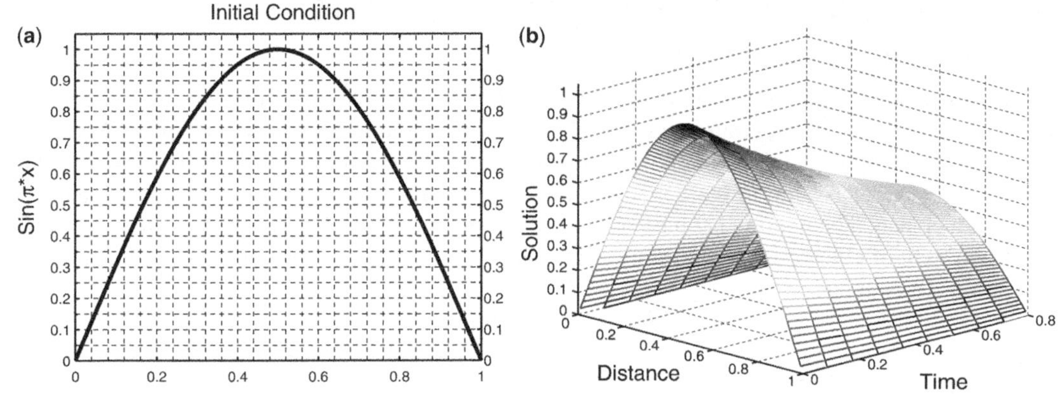

Figure 2.7 (a) Initial conditions and (b) solution for the diffusion equation.

When other types of boundary and initial conditions are given the solution has to be computed either through integration or through numerical approximation. In case of parabolic equations characteristics are not so important because they are equal, and they show that solution at point P of the computational space can influence any point in the computational domain. In case of parabolic equations any combination of Dirichlet, Neumann or Robin boundary conditions are appropriate to be used.

2.2.2.4 Elliptic equation example

Elliptic equations are time independent and therefore only valid in case of steady flows. A typical example of elliptic equation is the Laplace equation:

$$\frac{\partial^2 u}{\partial x^2} + \frac{\partial^2 u}{\partial y^2} = 0 \tag{2.37}$$

In case of elliptic PDEs all boundary conditions should be given (see Figure 2.8). The characteristic lines are of complex nature and cannot be represented in the real domain, hence they are not used. The important characteristic of an elliptic equation is that any disturbance at point P in the computational domain will influence all the points in the computational domain.

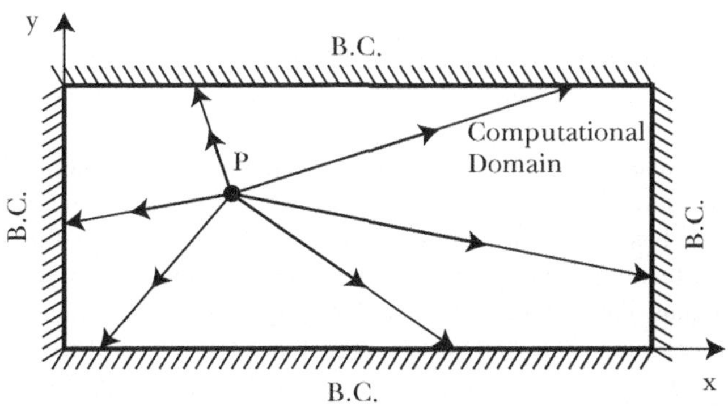

Figure 2.8 Computational domain and boundary conditions for elliptic PDEs.

2.3 NAVIER-STOKES AND SAINT-VENANT EQUATIONS

2.3.1 Navier-Stokes equations

The dynamics of fluid flow is governed by the Navier-Stokes equations, which are derived from the conservation laws (mass, momentum and energy), as shown in Section 2.1 of this chapter. Derivation of Navier-Stokes is shown in detail in many text books (Hirsch, 1988; Popescu, 2013). In case that flow of water is studied, several assumptions are made on the fluid; it is incompressible; viscous; homogenous; and does not conduct heat. The obtained equations are called incompressible Navier-Stokes equations and are written as a system of equations of continuity and momentum (Popescu, 2013). In case of incompressible flows equations (2.8) and (2.15) are:

$$\begin{cases} \dfrac{\partial \vec{u}}{\partial t} = -(\vec{u}\cdot\nabla)\vec{u} + \vartheta\nabla^2\vec{u} - \dfrac{1}{d}\nabla p + \vec{S} \\ \nabla\vec{u} = 0 \end{cases} \tag{2.38}$$

where $\vec{u}$ is the velocity vector, ϑ viscosity (controls the velocity of diffusion), p is pressure and $\vec{S}$ the external forces. First equation of the system (2.38) shows the change in velocity, which is composed of a summation of four terms. The first term in the right-hand side of the first equation of system (2.38) is called advection (convective) term, and it shows how the fluid moves around, it shows the force exerted on a particle of the fluid by the other particles of fluid surrounding it. The second term in the same equation is the diffusion term and it shows how the fluid motion is damped. The third term is the pressure term, and the fourth term are the external forcing functions such as gravity or wind.

Navier-Stokes equations are not easy to solve, because of the non-linearity of the advective term of the equation. Analytical solutions however, are available for particular cases, very simple ones that have a simple geometrical computational domain and simple boundary conditions.

For flow of water the Navier-Stokes equations are solved, by making further assumptions, which simplify the equations and allows for numerical approximation of the solution. Each of the simplified approach is used for different problems, because the assumptions made are depending on the considered phenomena. There are three main approaches, in water related flows, in order to represent Navier-Stokes equations in a simpler form:

- Averaging the equations: Reynolds averaged equations plus a turbulence model (RANS);
- Large Eddy Simulation (LES);
- Approximations of the averaged equations, that is, equations averaged over the cross-section (Depth averaged).

In the first approach, the average of the flow is used, and the turbulence fluctuations are considered less important. Each physical quantity is represented as an average over the computational time interval. The quantity itself is represented as a summation between a mean value and a fluctuating part. The averaging transforms the non-linear terms in linear ones, but they do introduce a new term, the derivative of the averaged quantity. The equations obtained are called Reynolds averaged equations. The additional term in the equation requires the use of a turbulence model. The LES forms of the Navier-Stokes equations are used for turbulence modelling (Aldama, 1990; Lesieur, 1997; Sagaut, 2001). In the third approach depth averaged approximations of velocity in space are used for the simplification of the Navier-Stokes equations.

2.3.2 Saint-Venant equations

The flow of water can be in a closed or open space. These two types of flow differ in the fact that in the open space water flows with a free surface, while in the closed space is a pressurised flow. The flow in a closed space is flow in conduits (pipes) and can be in some instances with free surface, however they can be considered the same as the flow in an open space. The free surface flow is the flow occurring in oceans, estuaries, coastal regions, lakes and impoundments. This kind of flow is subject to atmospheric pressure and water surface changes in space and time, hence it is more difficult to solve. Variables that are part of the equation; depth of flow, velocity or discharge, bottom slope, position of the free surface flow; are all interrelated. Moreover in case that flow is in a river the shape of the cross-section of the channel flow is arbitrary, while in pipe flow cross-sections are simple (circular or regular shape). Roughness of a channel may vary much more than the roughness of a pipe, which makes the free surface flow more complex. These equations are not easy to solve, hence research on methods of solution of the free surface flow equations has received considerable attention.

Due the complexity and importance of free surface flow, the equations are presented and analysed in detail here. Methods of solution for these equations through numerical approximations are shown in Chapters 4 and 5. For the special case of flow in rivers and lakes these equations and their solution is addressed in Chapters 7 and 8 of the book.

Free surface flow is unsteady in nature and the equations expressing it are unsteady flow equations. There are however cases in which the problems are steady in nature and the solution is easier to be determined. Though the equations of steady flow are determined separately, through assumptions on the steady character of the flow, they can also be determined from the unsteady flow equations, as particular case, when all changes with time are set to zero.

Flow is classified in respect with its variation with space, not only with time, as uniform and non-uniform flow. Flow is considered uniform if depth and velocity of flow is the same at every point in space. It occurs when gravity forces are in equilibrium with friction forces.

In case that the wave length of the free surface flow is much more bigger than the depth of flow then the equations describing the phenomena are referred as *shallow water flow equations* or *Saint-Venant* equations. These equations are Reynolds depth averaged forms of the Navier-Stokes equations. They can be determined also independently of Navier-Stokes equations, using physical arguments and by making assumptions on the fluid flow.

For simplicity of formulation Saint-Venant equations are further discussed for the case of one dimensional flow. The basic assumptions for the derivation of Saint-Venant equations are (Saint-Venant, 1871):

(1) The wave length of the disturbance of the flow is very long as compared with the depth of the flow. This assumption implies that flow is one dimensional and parallel to the walls and bottom forming a channel in which flows takes place;
(2) The channel geometry is not changing in time, that is, deposition or scour of sediment is very small and considered to have no effect;
(3) Streamline curvature of the channel is small; lateral and vertical accelerations are negligible compared to the longitudinal accelerations; therefore, the pressure distribution is hydrostatic;
(4) Slope of the channel bed is small, that is, the tangent and sine of the angle between bottom and horizontal have the same value as the angle itself;
(5) The effect of boundary friction force is estimated using a relation derived from steady uniform flow;
(6) Water surface in any cross section of the channel is assumed to be horizontal and parallel with the bottom of the channel.

There are two ways of expressing the Saint-Venant equations; in conservative form, when the dependent variables are discharge and depth of flow; and in non-conservative form, when velocities and water depth are the dependent variables. Equations expressed in velocities are non-conservative because velocities are not subject to conservation, and the equations do not hold across discontinuities, such as hydraulic jumps or in case of shocks. Equations in conservative form have the advantage that they do hold even if some of the assumptions of Saint-Venant equations break, as it is the case for example of the hydraulic jump. They are used when there are significant variations of flow in space.

The Saint-Venant equations are obtained by writing the conservation of mass and momentum over a small control volume. For a 1D channel, which has the longitudinal profile, and cross-section as the one represented in Figure 2.9, Saint-Venant equations for an arbitrary cross-section of a channel, in conservative form, are (Cunge *et al.*, 1986; Popescu, 2013):

$$\frac{\partial A}{\partial t} + \frac{\partial Q}{\partial x} = q \tag{2.39a}$$

$$\frac{\partial Q}{\partial t} + \frac{2Q}{A}\frac{\partial Q}{\partial x} + \left(-\frac{Q^2}{A^2} + \frac{gA}{B}\right)\frac{\partial A}{\partial x} - gAS_0 + gAS_f = qu_q \tag{2.39b}$$

|---| Kinematic wave
|--| Diffusion wave
|--| Dynamic wave

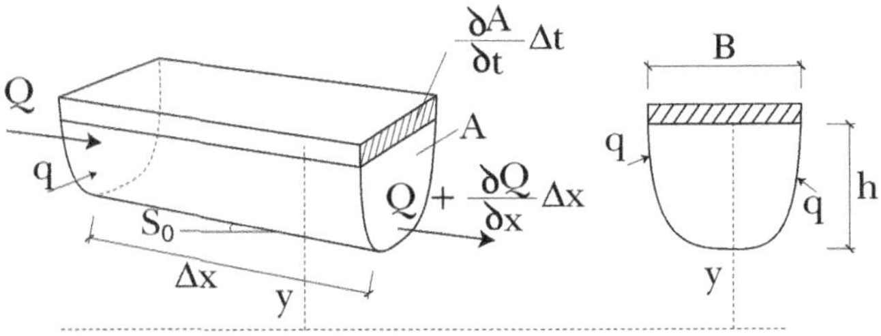

Figure 2.9 Control volume of a continuum fluid element.

Saint-Venant equations in non-conservative form in 1D are:

$$\frac{\partial h}{\partial t} + \frac{\partial (uh)}{\partial x} = q \tag{2.39c}$$

$$\frac{\partial u}{\partial t} + u\frac{\partial u}{\partial x} + g\frac{\partial h}{\partial x} - gS_0 + gAS_f = qu_q \tag{2.39d}$$

|---| Kinematic wave
|--| Diffusion wave
|--| Dynamic wave

In equations (2.39*a–d*) the notations are, as per Figure 2.9; A, area of the cross-section; Q discharge in the control volume; q lateral flow; h water depth; S_0 the slope of the channel, S_f the friction slope; u velocity of flow; and u_q velocity of the lateral flow.

Equation (2.39*a*) and (2.39*c*) represent the continuity equation, while (2.39*b*) and (2.39*d*) represent the momentum equation.

In both representations, momentum equation has five terms; the local acceleration, which describes the change of momentum due to the change of velocity over time; the convective acceleration that shows the change in momentum due to change of velocity along the channel; the pressure term that describes the pressure force which is proportional to the change in water depth along the channel; the gravity term that shows the force due to gravity; and the friction term that shows the friction force proportional to the friction slope.

The system formed by the above two equation, in conservative or non-conservative form, does not have an analytical solution. There are cases when momentum equation can be used with less than five terms, depending on the problem to be solved. Some of the terms, of the momentum equation might be very small, as compared to the other terms, hence they can be neglected (Price, 1985). Depending on how many terms are used in the representation of the equation of momentum, the following terminology is used:

- Kinematic wave representation, when momentum equation reduces to two terms: $S_0 = S_f$;
- Diffusive wave approximation, when the first two terms of the momentum equation, called inertia terms, are neglected; and
- Fully dynamic wave, when the full momentum equation is considered.

Solution approaches for these three cases are detailed below.

2.3.2.1 Kinematic wave solution

In the case of a kinematic wave approximation, the acceleration and pressure terms in the momentum equation are neglected. Kinematic wave approximation has no terms varying in time, the terms in the momentum equation represent the steady uniform flow. However the flow itself is not steady, because the unsteadiness is taken into account by the continuity equation. In case that there is no lateral flow ($q_l = 0$), the system of equations (2.39) has an analytical solution.

The slope of the energy line is the same as the slope of the channel bed, which is equivalent with stating that discharge is a function of flow depth, $Q = Q(h(x))$, hence:

$$\frac{\partial Q}{\partial x} = \frac{\partial Q}{\partial h} \cdot \frac{\partial h}{\partial x} \tag{2.40}$$

Inserting (2.40) into (2.38) it yields:

$$\frac{\partial A}{\partial Q}\frac{\partial Q}{\partial t} + \frac{\partial Q}{\partial x} = \frac{\partial A}{\partial Q}\frac{\partial Q}{\partial t} + \frac{\partial Q}{\partial h} \cdot \frac{\partial h}{\partial x} = q \tag{2.41}$$

Equation (2.41) is a form of advection equation, with the wave celerity $c = \frac{\partial Q}{\partial A}$ Re-arranging (2.41), for the case of zero lateral flow, yields:

$$\frac{\partial A}{\partial Q} \cdot \frac{dQ}{dh} \cdot \frac{\partial h}{\partial t} + \frac{dQ}{dh} \cdot \frac{\partial h}{\partial x} = 0 \tag{2.42}$$

Hence:

$$\frac{\partial h}{\partial t} + c \cdot \frac{\partial h}{\partial x} = 0 \tag{2.43}$$

An expression for the kinematic wave celerity c may be obtained from one of the steady flow equations such as Chezy or Manning (Chow *et al.*, 1988). For example by using Manning equation, and a wide cross-section approximation, for which hydraulic radius is approximately the same as the water depth, the value of the wave celerity is:

$$c = \frac{2}{3}u \tag{2.44}$$

Equation (2.43) is the *kinematic wave* equation and is a linear advection equation if c is a constant. The advection equation with a positive velocity c will propagate just in the positive direction of the x axis, that is, towards downstream. In general, the wave velocity is a function of water depth $c = c(h)$. The bigger the flow depth, the bigger the value for c. For one-dimensional problems this results in a steepening of the wave front. Kinematic wave equation applies to situations where flows are on steep bed channels, but do not apply to backwater effects, nor to tidal flows. An extensive study on the kinematic wave can be found in Ponce and Simon (1977).

2.3.2.2 Diffusive wave solution

The kinematic wave model is suitable just for some problems as it is for example the case of translation of a flood wave. A second approximation of the momentum equation is the one in which the pressure term is taken into consideration as well, hence:

$$\frac{\partial h}{\partial x} = S_0 - S_0 \tag{2.45}$$

which is equivalent with stating that $Q = Q\left(h(x), \frac{\partial h}{\partial x}\right)$. Replacing this relation into equation (2.38*a*), and considering lateral flow to be zero, lead to:

$$\frac{\partial Q}{\partial t} = c\frac{\partial Q}{\partial x} + D\frac{\partial^2 Q}{\partial x^2} \tag{2.46}$$

where c is the kinematic wave celerity (as previously defined) and D is the diffusion coefficient. Equation (2.46) has the form of an advection-diffusion equation, hence its name. The diffusion coefficient has the expression:

$$D = \frac{1}{B} \cdot \left| \frac{\partial Q}{\partial \left(\frac{\partial h}{\partial x}\right)} \right| \tag{2.47}$$

The diffusion coefficient is always positive because the energy slope is always positive. Different from kinematic wave, in case of diffusive wave, as the name is stating, there is diffusion. While the water wave travels towards the downstream direction diffusion in form of attenuation happens.

Applying the diffusive wave description of phenomena it is possible to solve problems with relatively small backwater effects, slowly propagating flood waves and runoff on mild slopes. Several methods of solution are available in the literature for the diffusive wave approximation such as Muskingum-Cunge method (Cunge, 1969), the Zero-Inertial solution (Strelkoff & Katopodes, 1977) or Parabolic and Backwater (Todini & Bossi, 1986).

2.3.2.3 Fully dynamic wave solution

In the case that inertial forces are as important as the pressure forces the fully dynamic expression of the momentum equation is used for the solution of Saint-Venant equations. Due to their complex nature these equations can not be solved analytically, however they can be solved using numerical methods, that is, using computer power to solve a system of equation obtained after the application of different numerical schemes to equations (2.38) or (2.39).

Popescu (2013) makes an overview of the different numerical techniques available for solving the Saint-Venant equations. These methods include the semi-analytical solution by the method of characteristics (Abbot & Basco, 1998), explicit difference methods, semi-implicit methods (Casulli, 1990), fully implicit methods, and Godunov methods (van Leer, 1979).

Solving Saint-Venant equations numerically implies that conditions on stability and convergence of the numerical solution must be met. Errors are mostly introduced by explicit methods and are avoided by imposing the Courant Frederichs Levy (C.F.L.) condition ($\Delta t \leq \Delta x/u$). Implicit methods do not induce stability problems, but are subject to constraints in the selection of the time and space step of computation.

2.3.3 Characteristic form of Saint-Venant equations

Saint Venant equations, as they are formulated in (2.39), can be transformed into the so-called characteristic form, which is useful to understand their meaning. Along with the formulation (2.39), the auxiliary variable $c = \sqrt{gh}$ is introduced, for which the derivatives with respect to independent variables x and t are:

$$\frac{\partial h}{\partial t} = \frac{2c}{g} \cdot \frac{\partial c}{\partial t} \tag{2.48a}$$

$$\frac{\partial h}{\partial x} = \frac{2c}{g} \cdot \frac{\partial c}{\partial x} \tag{2.48b}$$

Equations (2.48) are substituted in (2.39*a*) and (2.39*b*), then the obtained equations are divided by c and multiplied by g. As a result, the following two equations are obtained:

$$2\frac{\partial c}{\partial t} + 2u \cdot \frac{\partial c}{\partial x} + c \cdot \frac{\partial u}{\partial x} = 0 \tag{2.49a}$$

$$\frac{\partial u}{\partial t} + u \cdot \frac{\partial u}{\partial x} + 2c \cdot \frac{\partial c}{\partial x} = 0 \tag{2.49b}$$

Adding (2.49*a*) and (2.49*b*) and then subtracting them, results in:

$$\frac{\partial(u+2c)}{\partial t} + (u+c)\cdot\frac{\partial(u+2c)}{\partial x} = 0 \tag{2.50a}$$

$$\frac{\partial(u-2c)}{\partial t} + (u-c)\cdot\frac{\partial(u-2c)}{\partial x} = 0 \tag{2.50b}$$

Recall that the total derivative of an arbitrary function $f(x, t)$ is:

$$\frac{df}{dt} = \frac{\partial f}{\partial t} + \frac{dx}{dt}\cdot\frac{\partial f}{\partial x} \tag{2.51}$$

If $\frac{dx}{dt} = u + c$, then $\frac{df}{dt} = 0$. This is similar with saying that:

$$\frac{d(u+2c)}{dt} = 0 \quad along \quad \frac{dx}{dt} = (u+c) \tag{2.52a}$$

or equivalent

$$J_{+} = (u+2c) = const \quad along \quad \frac{dx}{dt} = (u+c) \tag{2.52b}$$

Similarly:

$$\frac{d(u-2c)}{dt} = 0 \quad along\ line \quad \frac{dx}{dt} = (u-c) \tag{2.53a}$$

or equivalent

$$J_{-} = (u-2c) = const \quad along \quad \frac{dx}{dt} = (u-c) \tag{2.53b}$$

Equations (2.52*a*,*b*) and (2.53*a*,*b*) are the characteristic equations; J_{-}, J_{+} are called Riemann invariants. These equations are of hyperbolic type. The fact that along the characteristic a special condition is valid shows that characteristics can pass information on the state of the fluid. The auxiliary variable, which has been introduced, is called celerity, and has a special meaning, it is the celerity with which a disturbance is propagated in water. Cunge *et al.* (1986) shows the relation between characteristic lines and the boundary and initial data requirements for the solution of Saint-Venant equations. In Figure 2.10 it is shown that in the case of subcritical flow at the upstream boundary of the computational domain one characteristic arrives from inside the domain, carrying information from another point of the computational domain. The other characteristic comes from the outside of the computational domain, hence it has to be replaced by a boundary condition. Such a boundary condition is given usually in the form of water depth, water velocity or discharge-stage relationship. The same analysis applies at the downstream boundary of the computational domain. In Figure 2.10b the case of supercritical flow is shown. It can be seen that the problem can be solved if two boundary conditions are given at the upstream boundary. The main conclusion is that for every characteristic entering a computational domain, one boundary condition has to be given.

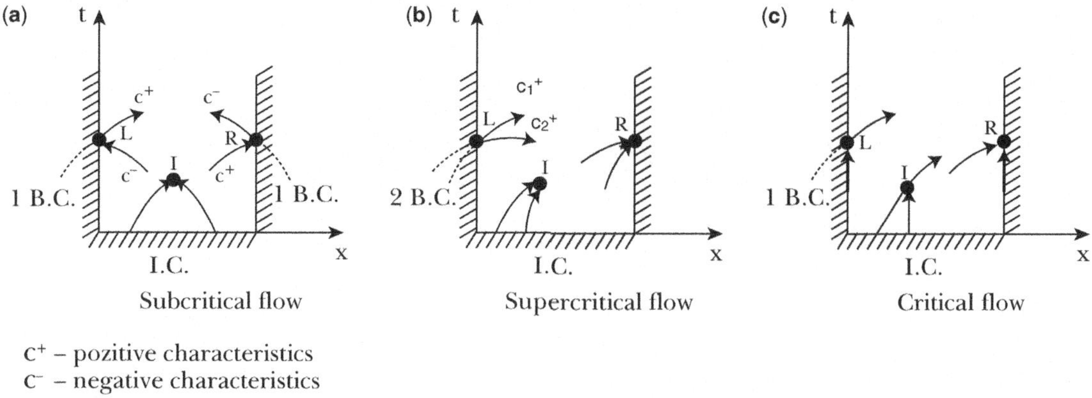

Figure 2.10 Characteristic solution of Saint-Venant equations over a given computational domain; (a) Subcriticalflow conditions; (b) Supercriticalflow conditions; and (c) Critical flow conditions.

Though this approach is very important, it cannot be applied to problems that include dissipative terms or for channels with irregular topographies. It is easier to use numerical approaches for finding the solution of Saint-Venant equations.

Chapter 3

Discretization of the fluid flow domain

3.1 DISCRETE SOLUTIONS OF EQUATIONS

In the previous chapter the engineering problems were represented by equations. Many of the equations do not have an analytical solution. There are two approaches to find the values of the unknown function of the equations; one is to make further assumptions such that the equations are simplified and brought to a form which has an analytical solution; and another is to develop approximate solutions using numerical methods.

The approach of simplifying the equations was successfully used in the past when computers were not available. The draw back of such an approach is that the solutions are valid just for a very limited number of simple engineering problems.

The approach of using numerical methods has been used and developed extensively as computer power grew with time. In order to apply the method of finding a solution for a differential equation through numerical approximation, the computational domain (space and time) has to be transformed from a continuum domain into a set of subdomains, where the equations are still valid. The procedure of transforming the space and time of the computational domain into smaller parts is called *discretization*. With such an approach the infinite number of degrees of freedom of continuous equations is transformed into a finite number, which can be solved using a computer, because the continuous unknown variables are evaluated on the subdomain as discrete values. Hence the smaller the discretization, the smaller is the approximation error.

There are different methods to discretize a computational domain, and in general the choice of the discretization method depends on the numerical method used for the solution of the equations. Each discretization method has its advantages and disadvantages.

The discretization of the computational domain in smaller parts, is done by a so-called *computational grid*, defined by nodes (or points). Nodes are defining cells. The discretization of the space domain implies that the governing equations of the phenomena are written in a discrete form, that is, a linear system of algebraic equations is obtained that would give the values of the unknown variables at grid nodes. The linear system of equation is solved simultaneously (with few exception when this is not needed, as is the case of the explicit approach).

In the process of transformation of the differential equations into a finite set of equations, through discretization, three stages are covered; the discretization of the definition domain of the differential equation; the discretization of the unknown function in values at discrete points (i.e., the approximate representation of the solution); and the discretization of the equation itself. Each of these stages is described separately in this book. Discretization of the computational domain is covered in the current chapter, while the discretization of the solution and of the equation are explained in the chapters about finite differences and finite volume methods. Regardless of the numerical method used, different types of grids are used in engineering problem practice. Figure 3.1 shows an example of a simple grid, representing the discretization of Yongdam reservoir, in South Korea. The gridlines cross each other orthogonally, hence the grid is called Cartesian.

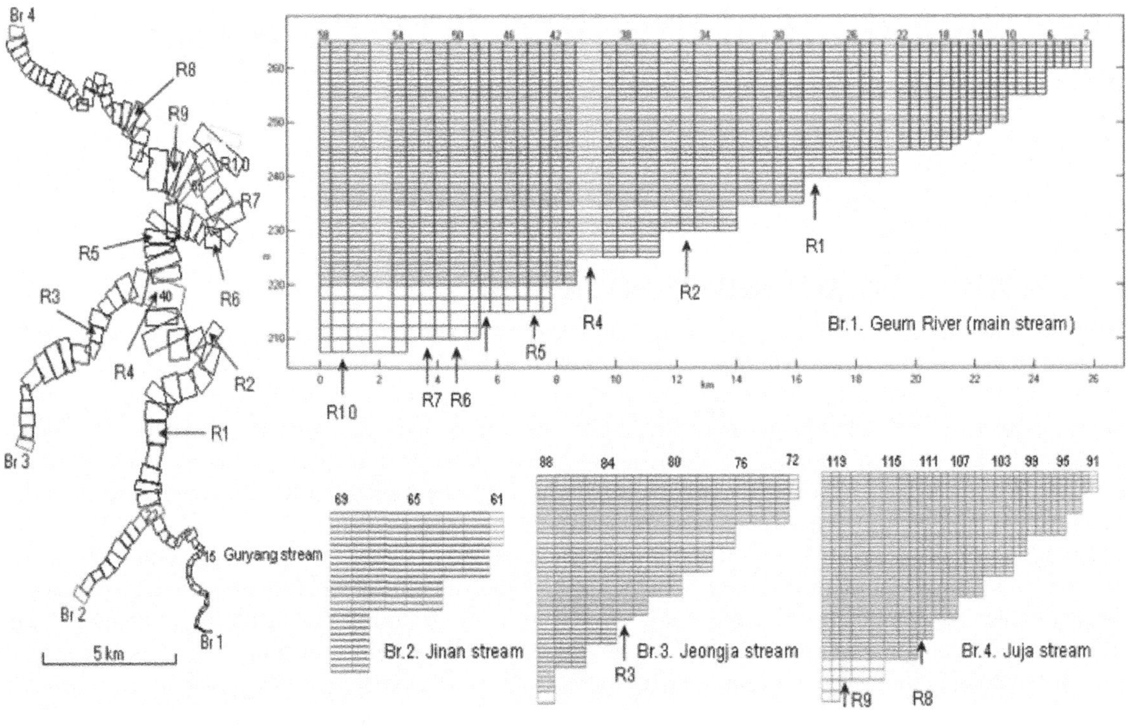

Figure 3.1 Vertical Cartesian computational grid, the example of Yongdam reservoir, South Korea.

The transformation of the continuous equations into discrete forms is done through methods such as finite difference, finite volume, finite element, boundary element, spectral methods, collocation methods, and so on. The first two are the methods used in this book and the presentation of grid generation is done in relation with these two methods. Discretization is done for both time and space, however the space discretization using in grids is more complex than that of time discretization, especially because it involves three independent variables (see Figure 3.2).

Initially space grids were regular and structured, however as more complex numerical methods were developed, due to the computational power of computers, irregular physical geometries could be accounted for. Space grids, also known as *mesh* are of three types; *structured, unstructured and meshless.* Meshless discretization is not used by the finite differences and finite volume methods therefore not addressed here.

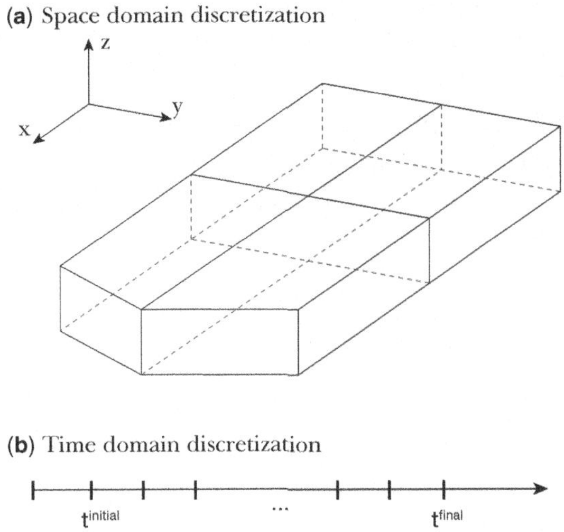

Figure 3.2 Conceptual discretization of the space and time domain.

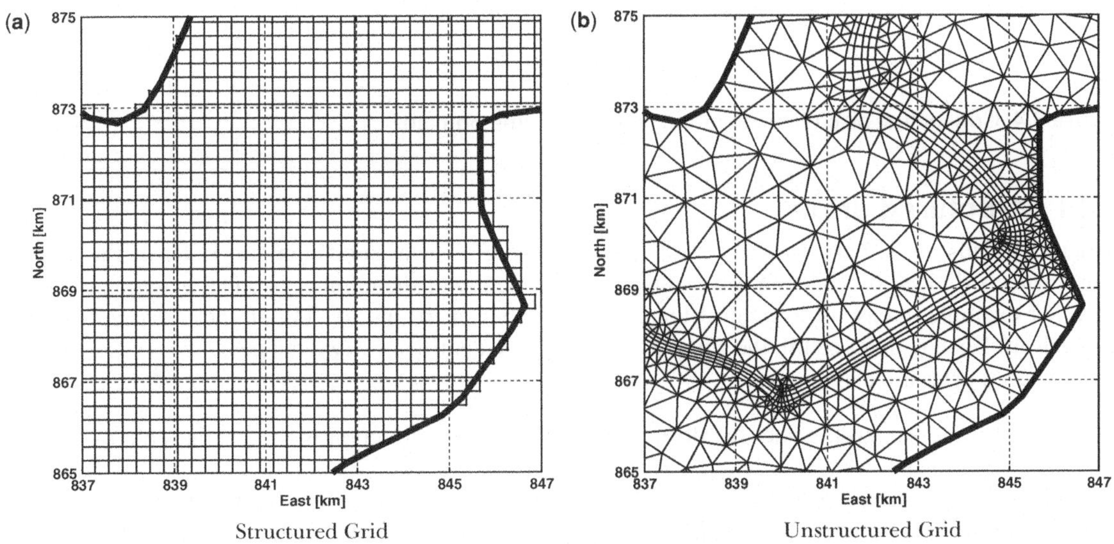

Figure 3.3 Example of structured and unstructured mesh.

The finite difference method requires a structured grid, while for the finite volume method both structured and unstructured meshes can be used (see Figure 3.3).

For each cell of a computational grid a space is allocated in the computer's memory, because the unknown function is defined at the nodes of the grid or at the cell points of the grid. In case the grid extends over domains that are not of interest, the required computational memory is very high.

Several author presents comprehensive reviews on how meshes are generated and for which type of problems they should be used (Buell & Bush, 1973; Thacker, 1980; Ho-Le, 1988).

3.2 SPACE DISCRETIZATION

Space discretization is discussed for two and three dimensions of space, because the one-dimensional space does not poses any challenge in discretization. One-dimensional computational domains are discretized in a computational grid by selecting nodes unequally or equally spaced (see Figure 3.4).

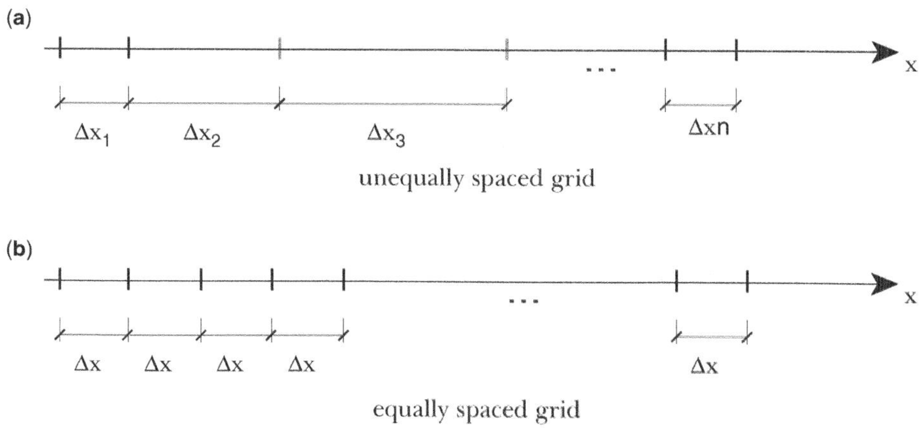

Figure 3.4 Conceptual discretization of the space and time domain.

3.2.1 Structured grids

Structured grids are simple, used for the discretization of a rectangular domain. Each of the axes of coordinates is divided by nodes in subdomains, independently of the number of subdivision on the other axes. For example in case of a 2D grid there are a number of nx and ny cells in the x and y direction, respectively. Each node or cell in the computational grid is identified by two indices.

Structured grids are classified as:

- Uniform regular grid, where the division of space in each direction is a constant (equally spaced) (see Figure 3.5a);
- Non-Uniform regular grid, where the division of space in each direction is not constant (see Figure 3.5b);
- Orthogonal grids, where the grid lines are orthogonal to each other (see Figure 3.5a and 3.5b);
- Curvilinear grids, where the grid lines are curves in space (see Figure 3.5c).

Regular structured grids are especially used with the finite difference method, because the derivatives of the unknown function are easily computed with respect to all spatial directions. Moreover these types of grids do not require a transformation of the coordinates, as is the case with the finite volume method, or other numerical methods. The use of such grids is easily implemented in a computer programme, because the nodes and cells are determined using their indices, hence the structure of the solution matrix is known in advance. The main disadvantage of structured grids is that they cannot represent irregular complex domains.

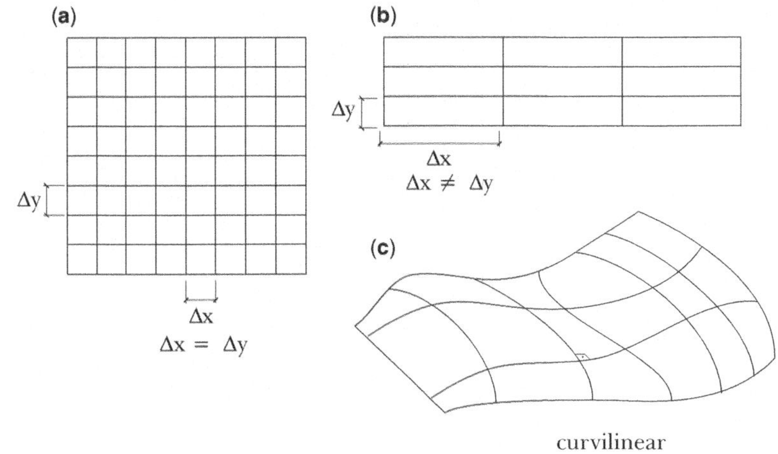

Figure 3.5 Examples of structured grids.

The only way for structured grids to partially cover irregular domains is by using the so-called multiblock grids, which are a combination of several structured grid blocks (see Figure 3.6). The solution computation is done on each block separately, however the computation advance from one block to another requires special treatment.

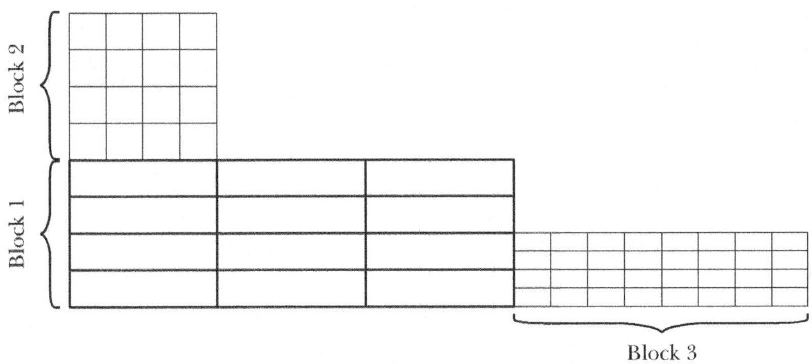

Figure 3.6 Example of a multi-block grid.

Apart from the multi-block type grids, which try to overcome irregular domains, there is another particular type of discretization grid called a *staggered grid*. These type of grids are specific to the problem that is solved by the differential equation and the way derivatives of the unknown function are expressed. They are used in case of a phenomenon that requires solutions for a system of differential equations and the different unknown variables have different approximation representations for the derivatives. For such grids different unknown variables are determined in different nodes of the grids.

An example of a staggered grid in one dimension is shown in Figure 3.7 in which Saint Venant equations are solved for a one-dimensional flow in open channels.

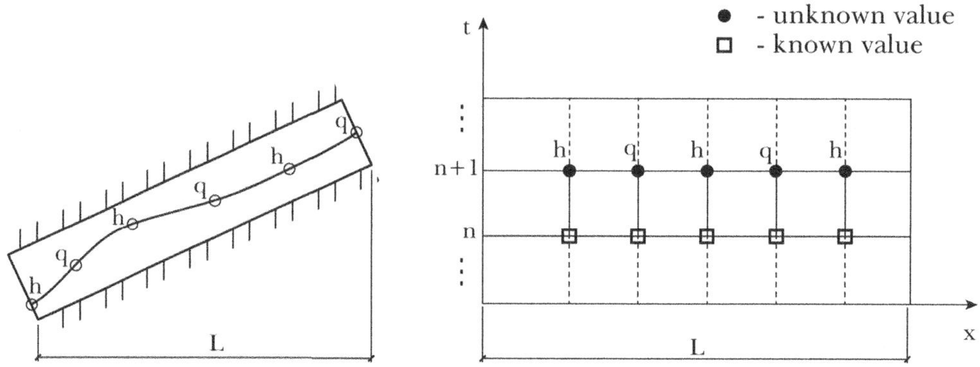

Figure 3.7 Example of a staggered grid.

3.2.2 Unstructured grids

Unstructured grids are complex and are not generated easily, however they are highly flexible in representing the computational domain. The number of nodes and cells of an unstructured grid are not related by an equation, hence it is not easy to identify a cell using indices only, as in the case of structured grids. Each cell of an unstructured grid has to be defined by the nodes that are forming it, therefore all coordinates of the nodes have to be defined along with the information regarding the cell to which they belong. Unstructured grids are used for the finite volume method and they are the grid of preference for the finite element method. They have the advantage that they can represent some areas of the computational domain in a very detailed manner, however they have the disadvantage of longer computational time. An example of an unstructured grid is shown in Figure 3.8, as general representation. Figure 3.9 shows the structured and unstructured grid representation for a portion of Yellow River, in China.

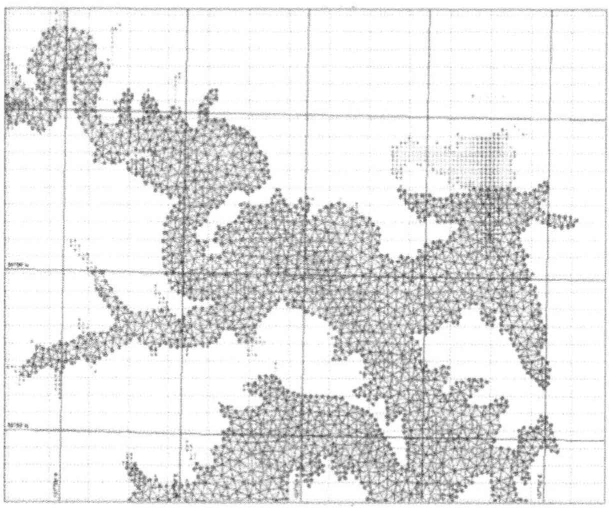

Figure 3.8 Example of unstructured grid.

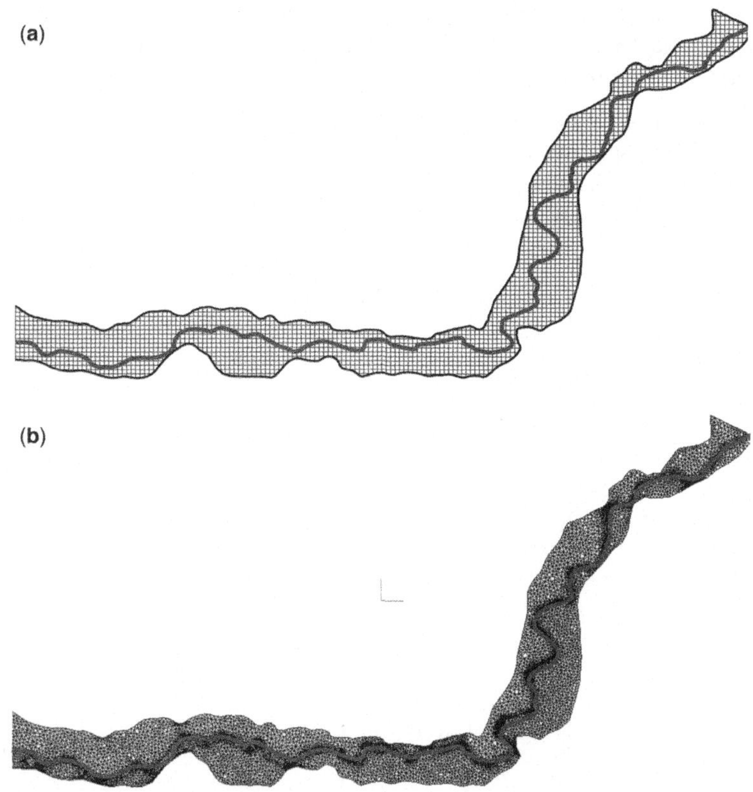

Figure 3.9 Structured (a) and unstructured (b) grid representation for the middle section of the Yellow River, China.

3.2.3 Grid generation

Grid generation is a difficult task, and can be performed manually just for simple domains that require a small number of cells. For complex computational and/or large computational domains tools are used to create the grid of a given domain. Such tools are called *grid generators*. Grid generation is a field of study on its own and several methods are available to develop grids. These methods use algebraic relations to create the nodes and cells of a grid. Grid generators can take into account several requirements to a grid such as maximization of the cell area or orthogonality. In general grid generators are dependent on the type of differential equation to be solved, that is, the grid generators for hyperbolic equations are different from the elliptic ones, because of the type of boundary conditions that need to be defined at the nodes of the grid.

In general grid generators produce a computational domain covered by a grid of triangles, by first creating all the nodes of the grid and then connecting them to form triangles. The best way to form these triangles is defined by the Delaunay triangulation, which ensures that very thin triangles are avoided (Boubez *et al.*, 1986); or by using the Voronoi diagram (Brostow & Dussault, 1978). The Voronoi diagrams consist in building N polygons *V*(*i*) out of N given points, in such a way that each polygon is centered in point *i*, and the locus of the points most closest to point *i* is included in *V*(*i*).

The output of a grid generation is a set of elements along with a set of points that define it. The set of elements represents the *grid topology*. Depending on the approach taken to create the elements of a grid, the grid generators are (see Figure 3.10):

- Creating first nodes then elements;
- Creating first grid topology then nodes;
- Creating nodes and elements simultaneously; and
- Adapting grid from pre-defined grid templates.

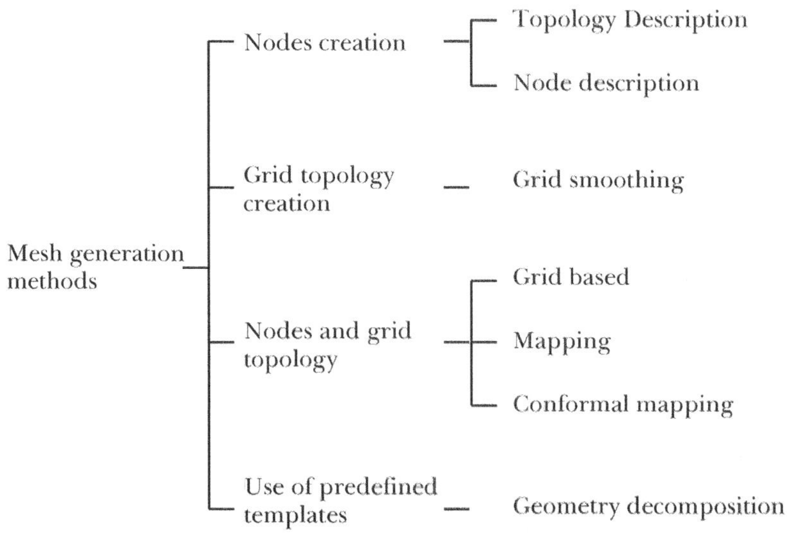

Figure 3.10 Classification of grid generation methods.

From the four methods listed above the templates are generated separately, independently of the computational domain and they are later adapted for the given computational domain. The approaches known to create templates are: grid-based, mapped elements, conformal mapping and nodes and elements created simultaneously.

There are several quality checks that need to be performed over a grid after it has been generated using a grid generator. The most important aspects are: grid smoothing, grid density and grid conformity. In case these three qualities of a grid are not fulfilled then the grid has to be refined manually.

Grid smoothing refers to the shape of the elements. Smoothing is performed by repositioning the nodes. Repositioning is a difficult task that requires iterations, as repositioning might be wrong.

Density of a grid refers to the ability of a grid to change from one part of the computational domain to another depending on the computational needs. Conformity is related to the density of a grid, because when a grid is refined smaller elements are introduced and they might be distorted. In general orthogonality is the measure to check for grid conformity.

3.2.4 Physical aspects of space discretization

There are problems for which the domain over which the solution is sought is bounded by special elements, such as structures, or may have a special form in the physical plane. Such special domains with special boundaries may be discretized to obtain the computational domain, which may be represented in a simpler format through a simple transformation of coordinates. After the solution is computed all values of the unknown function are placed back to the initial computational domain, which represents the real domain. Such an example is a distorted L-shape region that can be transformed into a simple I-shape (see Figure 3.11).

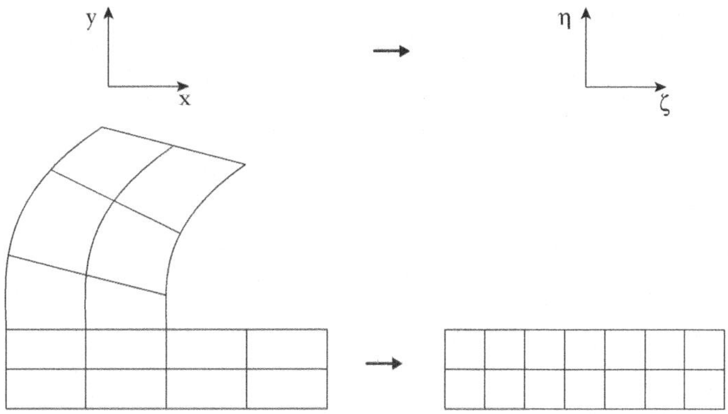

Figure 3.11 Computational domain transformations from one system of coordinates to another.

Transformation should be done with care because it may introduce accuracy problems in the solution. In case of a bump a transformation grid may eliminate the bump (see Figure 3.12).

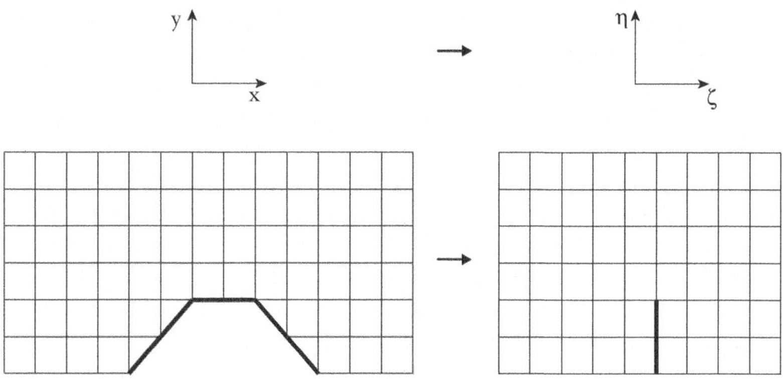

Figure 3.12 Computational domain transformation for a bump.

Fletcher (1991) emphasizes the importance of carefully making such transformations because the computational domain is distorted and so is the corresponding approximate solution.

3.3 TIME DISCRETIZATION

Discretization of time is separated from the one in space. Time discretization is used for problems, where the boundary of the computational domain varies in time, as is the case with free surface flows. In each time step values of the unknown function are computed at all nodes of the grid in space. When all values of the unknown function have been determined at time step n computation advances to the next time step

$(n + 1)$. Numerical methods algorithms are written with respect to time level n and $(n + 1)$, considering that all values of the unknown function are known at time step n, for all nodes of the space grid, and a new computation starts to determine all the values at time level $(n + 1)$ where values of the unknown function are considered unknown (see Figure 3.13).

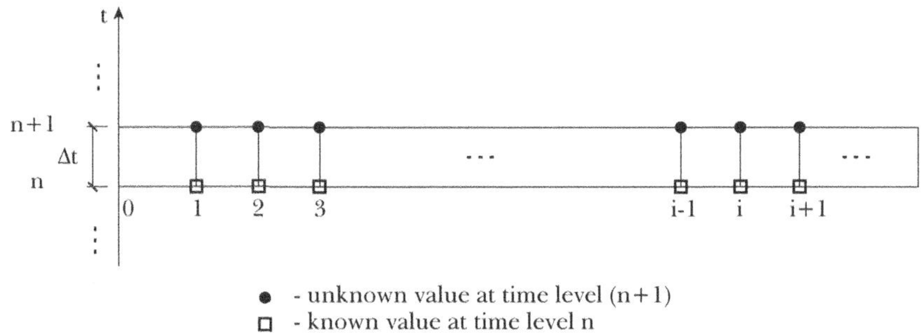

Figure 3.13 Time discretization.

Depending on the numerical method to determine the approximate solution of a differential equation, there are two main types of discretization with respect to time; explicit and implicit methods.

In explicit discretization the values at the grid nodes of each unknown variable at the new time step of computation depend only on known values at the previous time step. The advantage of this kind of discretization is that in many cases no system of equation is formed, hence no solver is required, because the solution is straightforward computed step by step. Such a discretization is easy to implement in a computer programme. The disadvantage of such an approach is that solution becomes unstable if the time intervals for computation are selected bigger than a threshold value. Such a threshold value is usually determined from stability conditions of the solution combined with physical characteristics of the variable that needs to be determined.

The implicit method of time discretization determines the values of the unknown function at the grid nodes from values of the function at neighbouring points at the same time level. An algebraic system of equations is formed, that requires a solver to be solved. The advantage of such method is that they are stable and no threshold value on the time interval of computation is required. From an accuracy point of view one should always try to solve the problems with small time intervals.

In order to take advantage of both implicit and explicit methods of discretization for time, combined methods are also used. They are called semi-implicit methods, where the variable at the new time step depends both on the old and new time step.

Chapter 4
Finite difference method

4.1 GENERAL CONCEPTS

The finite difference method (FDM) dominated computational science since its beginnings. This was the method of choice in computational hydraulics in the 1960s and 1970s. While other methods, such as the finite element method and the boundary element method have been extensively used, the numerical schemes developed using the finite difference approach are still used for a wide range of engineering problems.

The finite difference method uses the differential form of equations representing a physical phenomenon, and solves the system of equations by determining values of the unknown variable function in discrete nodes of the definition domain. The domain is discretised in grids, as shown in Chapter 3. The values of the unknown variable are determined in the nodes of the grid. The value of the variable in any point of the computational domain, that is not a discretization node, is obtained by interpolation between the known values at discretization nodes.

The finite difference method can be characterized as follows:

- Utilizes spaced grids, mainly uniform ones;
- At each node, each derivative is approximated by an algebraic expression that relates the values of the unknown function of the variable with reference to the adjacent nodes;
- A system of algebraic equations is obtained and solved for the unknown variable, which is also known as dependent variable.

The applied finite difference method is expected to generate estimates of the unknown function that will reproduce the analytical shape of the same function with sufficient accuracy. The accuracy depends on the selected nodes (points) of discretization and the calculation method of the unknown function derivative.

In case time is one of the independent variable of the equation to be solved, three kinds of numerical schemes are used in practice:

- Schemes in which the value of the unknown variable at time level $n + 1$ can be computed directly (or explicitly) from the value at time level n. Such schemes are called ***explicit*** schemes.

- Schemes in which the determination of the derivative involves the value of the unknown variable at time level $n + 1$. In such schemes there is an implicit relationship between the derivative and the unknown variable, hence they are called ***implicit*** schemes.
- Schemes in which the explicit and implicit approaches are ***combined***.

Ordinary differential equations, depending on the computational approach of the numerical scheme, are classified as ***one-step*** methods and ***multi-step*** methods. If the unknown function at a node of the grid can be computed using information only from the previous node of the discretization grid then the method is called one-step. One-step methods are also known as self starting methods, because the given initial condition is sufficient to perform the computation for the value of the unknown function in all the nodes of the discretization grid. If the value of the unknown function in the current computational step is computed based on information from more than one preceding step of computation then the method is called multi-step.

Each type of scheme has very different properties, as shown in Chapter 6 of the book. There are many finite difference methods available for solving differential equations, however due to the properties of the equations used in the aquatic environment, only few of them are applicable, therefore only a selection of these methods are presented in this chapter.

4.2 APPROXIMATION OF THE FIRST ORDER DERRIVATIVE

Given a function $f(x)$ with one independent variable x, the definition of its differential with respect to x, as given in elementary calculus, is (see Figure 4.1):

$$f'(x) = \lim_{\Delta x \to 0} \frac{f(x + \Delta x) - f(x)}{\Delta x} \tag{4.1}$$

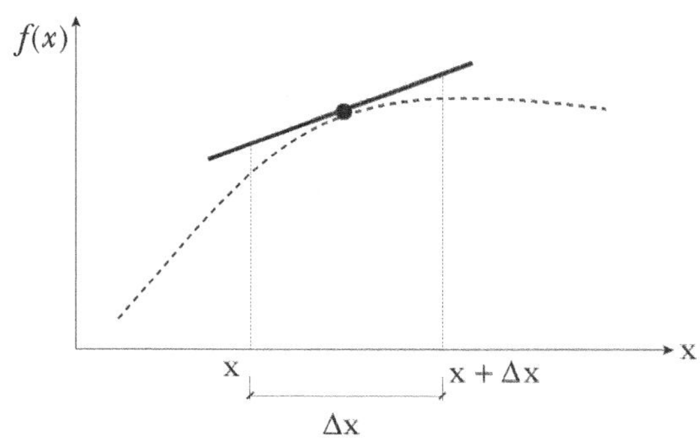

Figure 4.1 Definition of the differential.

Nowadays calculations are carried out on computers and these cannot deal with a limit for which $\Delta x \to 0$, therefore a discrete form of the continuous case is introduced.

The function $f(x)$ is defined by a set of values available at a number of discrete points of the domain of definition of the function. Figure 4.2 shows a set of discrete points x_i where the function is known. The nodes x_i are positioned at distance intervals $\Delta x_i = x_{i+1} - x_i$. The notation f_i is used to denote the value of the function $f(x_i)$ at the i-th node, that is, $f_i = f(x_i)$. The set of nodes is called the computational grid (as already defined in Chapter 3).

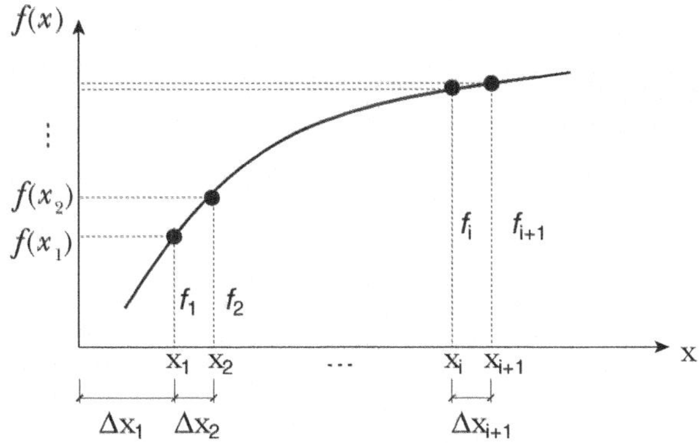

Figure 4.2 Discretization nodes for a one-dimensional function.

The approximation of the function's derivative is done using the discrete values of function f in a number of points of the domain of the function. This approximation is obtained using Taylor series expansions.

Consider, for simplicity, an equally spaced grid of the independent variable x. A Taylor series is an expansion of $f(x + \Delta x)$ or $f(x - \Delta x)$ for a finite Δx at a fixed node x and can be expressed as:

$$f(x+\Delta x) = f(x) + \Delta x f'(x) + \frac{(\Delta x)^2}{2!} f''(x) + \frac{(\Delta x)^3}{3!} f'''(x) \ldots \tag{4.2}$$

$$f(x-\Delta x) = f(x) - \Delta x f'(x) + \frac{(\Delta x)^2}{2!} f''(x) - \frac{(\Delta x)^3}{3!} f'''(x) \ldots \tag{4.3}$$

where f' is the first derivative of f with respect to x, f'', the second derivative and so on.

Rearranging (4.2) gives:

$$\frac{f(x+\Delta x) - f(x)}{\Delta x} = \frac{1}{\Delta x}\left[\Delta x f'(x) + \frac{(\Delta x)^2}{2!} f''(x) + \frac{(\Delta x)^3}{3!} f'''(x) + \cdots\right] = f'(x) + O(\Delta x) \tag{4.4}$$

If equations (4.2) and (4.3) are subtracted from each other it yields:

$$\frac{f(x+\Delta x) - f(x-\Delta x)}{2\Delta x} = \frac{1}{2\Delta x}\left[2\Delta x f'(x) + 2\frac{\Delta x^3}{3!} f'''(x) + \cdots\right] = f'(x) + O(\Delta x^2) \tag{4.5}$$

If Δx is small then it can be considered that:

$$f'(x) \approx= \frac{f(x+\Delta x) - f(x)}{\Delta x} \tag{4.6}$$

or

$$f'(x) \approx= \frac{f(x) - f(x-\Delta x)}{\Delta x} \tag{4.7}$$

or

$$f'(x) \approx= \frac{f(x+\Delta x) - f(x-\Delta x)}{2\Delta x} \tag{4.8}$$

Equations (4.6)–(4.8) are examples of three simple finite difference approximations, as represented in Figure 4.3; the *forward, backward* and *centered* difference, respectively.

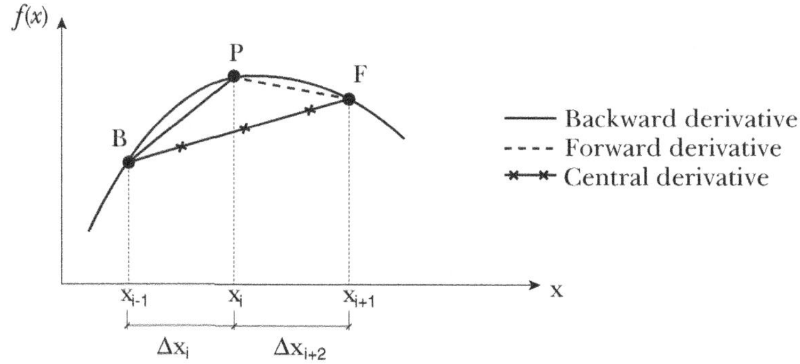

Figure 4.3 Forward, backward and central approximations of the derivatives.

In Figure 4.3 function $f(x)$ is represented for a small interval near the discretization point x_i. It can be seen that forward differencing uses the forward values of function $f(x)$, that is, the values of the function in point $(x_{i+1}, f(x_i + \Delta x_i))$, backward values in point $(x_{i-1}, f(x_i - \Delta x_{i-1})$ for backward differencing and values on both sides of the function, in points $(x_{i+1}, f(x_i + \Delta x_i))$ and $(x_{i-1}, f(x_i - \Delta x_{i-1}))$ for central differencing.

The term $O(\Delta x)$ in equation (4.4) or $O(\Delta x^2)$ in equation (4.5) is a truncation of the Taylor series after the second term (right hand side of the equation). These terms are referred to as the truncation error (T.E.) because they represent the error when equations (4.6)–(4.8) are used. Hence the T.E. can be defined as the difference between the partial derivative and its finite difference representation.

For smooth functions (the ones that have continuous higher order derivatives), and for a sufficiently small Δx, the first term of the truncation error, can give the order of magnitude of the error. Depending on the order of the term $O(\Delta x)$, the following terminology is used:

- Equation (4.3) is first order accurate in Δx
- Equation (4.4) is second order accurate in Δx
- Equation (4.6) is a higher order approximation to $f'(x)$ than (4.3) or (4.4), which shows that if $\Delta x \to 0$, (4.5) $\to f''(x)$ in a faster way than (4.4). It is important to note that the term is higher order not better approximation.

Several higher order approximations of the first derivative of a function can be obtained by linearly combining the two Taylor series representations given by equations (4.2) and (4.3).

In case that the discretization grid is not equally spaced (see Figure 4.2), equations (4.2) and (4.3) are:

$$f(x_i + \Delta x_i) = f(x_i) + \Delta x_i f'(x_i) + \frac{(\Delta x_i)^2}{2!} f''(x_i) + \frac{(\Delta x_i)^3}{3!} f'''(x_i) + \cdots \tag{4.9}$$

$$f(x_i - \Delta x_{i-1}) = f(x_i) - \Delta x_{i-1} f'(x_i) + \frac{(\Delta x_{i-1})^2}{2!} f''(x_i) - \frac{(\Delta x_{i-1})^3}{3!} f'''(x_i) + \cdots \tag{4.10}$$

If equation (4.9) multiplied by Δx_{i-1}, is added to equation (6.10) multiplied by Δx_i, the following equation is obtained:

$$\frac{1}{\Delta x_i - \Delta x_{i-1}}\left[\Delta x_{i-1}\frac{f(x_i + \Delta x_i) - f(x_i)}{\Delta x_i} + \Delta x_i \frac{f(x_i) - f(x_i - \Delta x_{i-1})}{\Delta x_i}\right] - f'(x_i) = \frac{\Delta x_i \Delta x_{i-1}}{3!} f''(x_i) + \cdots \tag{4.11}$$

It can be noticed that in case of an equally spaced grid equation (4.11) is simplified to equation (4.6), that is, the cantered difference approximation is obtained.

From equation (4.11), it can be noticed, that three values of the unknown function in three discretization nodes are used to compute the discrete value of the centered derivative at the node x_i, that is, the nodes x_{i-1}, x_i and x_{i+1}. The truncation error $O(\Delta x_i \Delta x_{i-1})$ is of second order.

Equation (4.11) can be used to determine higher order approximations for the first derivative of a function. For an equally spaced grid, the following expression holds:

$$\frac{8(f(x_i + \Delta x) - f(x_i - \Delta x))}{12\Delta x} = f'(x_i) - \frac{4(\Delta x)^4}{5!} f^{(5)}(x_i) - \frac{20(\Delta x)^6}{7!} f^{(7)}(x_i)\ldots \tag{4.12}$$

If equation (4.11) multiplied by 4 is subtracted from equation (4.11), written for a $2\Delta x$ grid spacing, equation (4.12) is obtained, which is a fourth order approximation of $f'(x)$.

It is important to be noted that the validity of the Taylor series regarding the existence of higher order approximations of the derivatives depends on the existence of the derivatives themselves. If derivatives do not exist, then the higher order approximations do not hold.

In a similar way the first derivative of a function of one variable is approximated by a finite difference, a first order partial derivative of a function depending on two or more variables can be approximated by finite differences.

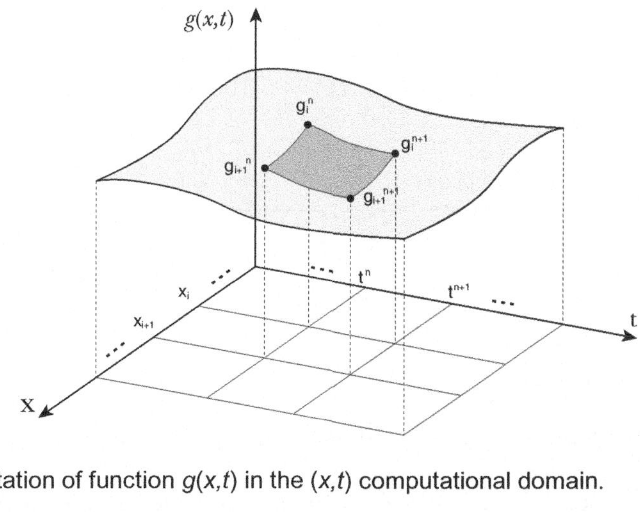

Figure 4.4 Representation of function $g(x,t)$ in the (x,t) computational domain.

Let $g = g(x, t)$, than the first order partial derivatives of the function with respect to x or t are (see Figure 4.4):

$$\left.\frac{\partial g}{\partial x}\right|_{(i,n)} \approx = \frac{g_{i+1}^{n} - g_{i}^{n}}{\Delta x} \tag{4.13}$$

and

$$\left.\frac{\partial g}{\partial t}\right|_{(i,n)} \approx = \frac{g_{i}^{n+1} - g_{i}^{n}}{\Delta t} \tag{4.14}$$

where the following notations were made:

- $\left.\frac{\partial g}{\partial x}\right|_{(i,n)}$ and $\left.\frac{\partial g}{\partial t}\right|_{(i,n)}$ are the partial derivatives of function $g(x, t)$ at node $(x_i; t^n)$, with respect to independent variables x and t, respectively;
- g_i^n is the value of the function $g(x, t)$ at node $(x_i; t^n)$;
- subscript indices are used for space, while superscript indices are used for time.

4.3 APPROXIMATION OF HIGHER ORDER DERRIVATIVES

Differential equations of phenomena in water related sciences contain higher order derivatives. These are expressed using different approaches, such as polynomial fitting or Taylor series expansion, which provide a systematic way to estimate higher order derivatives of any order. It is important to note that the function for which the derivatives are approximated should be smooth.

The approximation of higher order derivatives using Taylor series expansion is demonstrated here. The function is considered to vary with several independent variables, therefore partial derivatives are used to derive the expressions. The independent variable is considered to be x.

Consider an equally spaced grid on which higher order derivatives are approximated in the node i (point of interest). A number of l points are considered at the *left* side of nodes i, and r at the *right* side of the node (see Figure 4.5). The value of function u in grid point $i + m$ $(m = -l, \ldots, r)$ based on value of the function in grid point i; and the higher order partial derivatives in the same point, is given by:

$$u_{i+m} = u_i + \frac{m\Delta x}{1!}\frac{\partial u}{\partial x} + \frac{(m\Delta x)^2}{2!}\frac{\partial^2 u}{\partial x^2} + \frac{(m\Delta x)^3}{3!}\frac{\partial^3 u}{\partial x^3} + \frac{(m\Delta x)^4}{1!}\frac{\partial^4 u}{\partial x^4} + \cdots \tag{4.15}$$

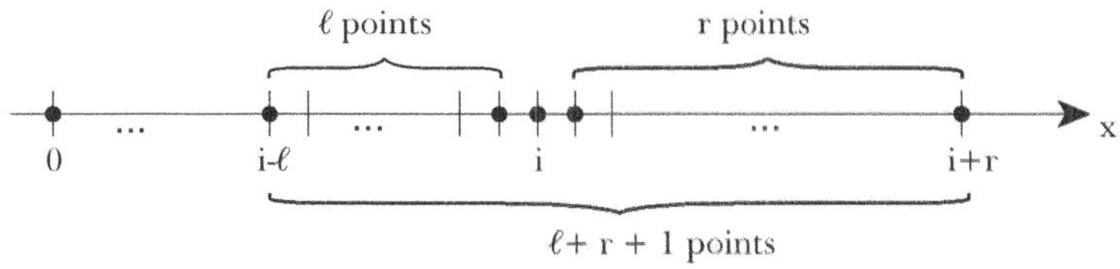

Figure 4.5 Equally spaced grid on the x axis.

In equation (4.15), time level n has been fixed, hence the indices for the independent variables t are not shown, for simplicity of expression.

If each of the $r + l + 1$ equations of type (4.15) is multiplied by a constant coefficient a_m and then all are summed up:

$$\sum_{m=-l,\, m\neq 0}^{r} a_m \cdot u_{i-m} - \left(\sum_{m=-l,\, m\neq 0}^{r} a_m\right) u_i = \left(\sum_{m=-l,\, m\neq 0}^{r} m \cdot a_m\right) \frac{\Delta x}{1!} \frac{\partial u}{\partial x}\bigg|_i + \left(\sum_{m=-l,\, m\neq 0}^{r} m^2 \cdot a_m\right) \frac{(\Delta x)^2}{2!} \frac{\partial^2 u}{\partial x^2}\bigg|_i$$
$$+ \left(\sum_{m=-l,\, m\neq 0}^{r} m^3 \cdot a_m\right) \frac{(\Delta x)^3}{3!} \frac{\partial^3 u}{\partial x^3}\bigg|_i + \left(\sum_{m=-l,\, m\neq 0}^{r} m^2 \cdot a_m\right) \frac{(\Delta x)^4}{4!} \frac{\partial^4 u}{\partial x^4}\bigg|_i + \cdots \tag{4.16}$$

Equation (4.16) is used to determine the $r + l$ coefficients a_m of the derivative depending on the desired order. For example if the numerical solution requires a first order derivative, fourth order accurate the values of the coefficients b_i are set as follows:

$$\begin{cases} b_1 = 1 \\ b_2 = 0 \\ b_3 = 0 \\ b_4 = 0 \end{cases} \tag{4.17}$$

where

$$b_k = \sum_{m=-l,\, m}^{r} m^k a_m \tag{4.18}$$

The system (4.17) gives four equations that would require the use of four discretization points, in order to have a unique solution of the system.

In general if the k-th derivative is needed then the highest derivative to be neglected is of order $k + p - 1$, that is, $k + p - 1$ grid points are needed to determine the solution, and equation (4.18) can be re-written as:

$$b_q = \delta_{qk} = \sum_{m=-l,\, m}^{r} m^q a_m; \quad q = 1, 2, \ldots, k + p - 1 \tag{4.19}$$

with

$$\delta_{qk} = \begin{cases} 1, & \text{if } q = k \\ 0, & \text{if } q \neq k \end{cases} \tag{4.20}$$

with δ_{qk} known as the Kronecker delta. The solution exist and it is unique if:

$$l + r = k + p \tag{4.21}$$

Based on (4.19)–(4.21) the order of the T.E. is determined by multiplying the next higher derivative in the truncation error series with the coefficient:

$$b_{k+1} = \sum_{m=-l,m}^{r} m^{k+p} a_m \tag{4.22}$$

Based on (4.18), selecting the centred difference approximation, and considering an equally spaced grid, it can be noticed that the second order derivative of a function $u(x, t)$, for which the domain is represented in Figure 4.6, can be computed as:

$$\left.\frac{\partial^2 u}{\partial x^2}\right|_{(i,n)} \approx = \frac{u_{i+1}^n - 2u_i^n + 2u_{i-1}^n}{(\Delta x)^2} \tag{4.23}$$

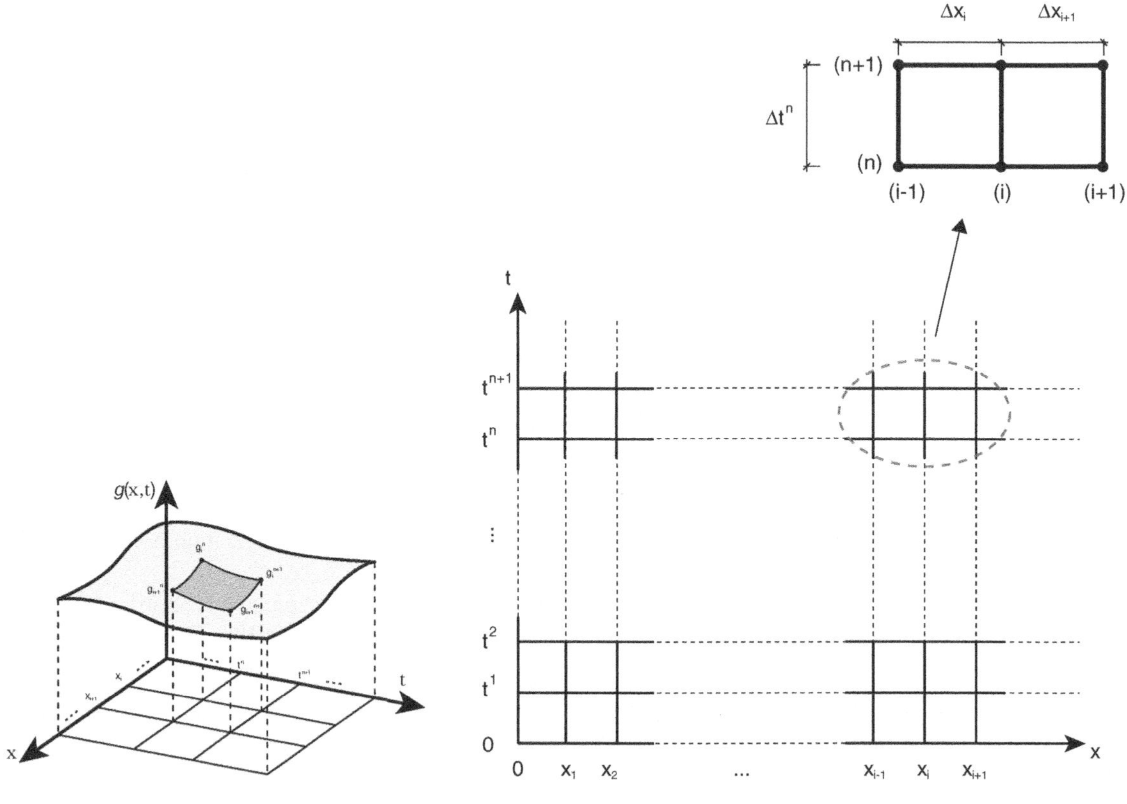

Figure 4.6 Domain of definition and grid discretization for function *u*(*x*, *t*).

Another way of developing finite difference approximations are, as previously mentioned, polynomial fitting, such as splines or Lagrange interpolations. These approaches consist in fitting a polynomial of a certain degree to a set of selected nodes of the grid. For example, in case of a function $u(x)$, of one independent variable, the slope of the function u, at node x_i is approximated by considering the derivative of the considered polynomial at the point (x_i, u_i). The development of these approaches are presented in detail in Boyd (1989), Butcher (1987) and Duran (1999).

4.4 FINITE DIFFERENCES FOR ORDINARY DIFFERENTIAL EQUATIONS

4.4.1 Problem position

Ordinary differential equations (ODEs) contain a function that varies with respect to one independent variable only. They contain derivatives of the unknown function and/or the function itself. The general form of an ODE is:

$$u' = \frac{du}{dt} = f(u,t) \tag{4.24}$$

where $u = u(t)$ is the unknown function, t is the independent variable and $f(u, t)$ is an arbitrary function of $u(t)$ and t.

Integrating equation (4.24) yields:

$$u(t) = \int f(u,t)\,dt + C \tag{4.25}$$

where C is a constant of integration. There are an infinite number of values for C, hence the obtained solution of equation (4.24) is not unique. However in the daily practice of engineering solving problem, there is interest just for a unique solution for a specified range of t, over a given interval. In order to choose a solution an extra condition, called *initial condition*, must be specified.

The problem is formulated as follows: Determine the unknown function $u(t)$, over the interval ($t^{initial}$, t^{final}) (see Figure 4.7), which satisfies the following set of equations (the given ODE and the initial condition):

$$\begin{cases} \dfrac{du}{dt} = f(u,t) \\ u(t^{initial}) = u^{initial} \end{cases} \tag{4.26}$$

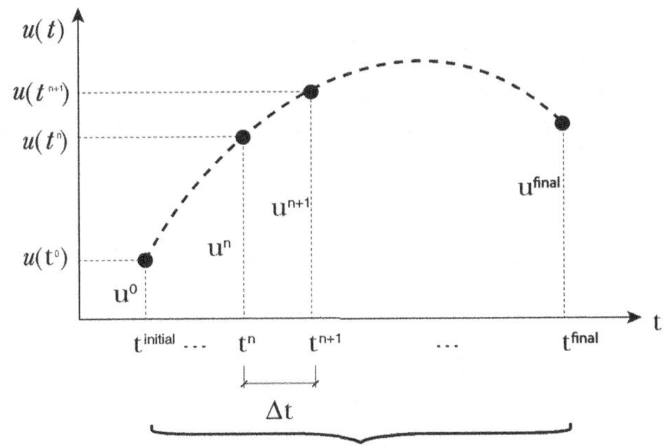

Figure 4.7 Discretization domain and solution approximation for function $u(t)$.

The problem formulated by (4.26) is called an *initial value* problem (IVP), for the ODE, and has a unique solution (Asher & Petzold, 1998).

Examples of engineering problems in water related field that are represented by equations like the ones of (4.26) are the steady flow in open channels; unit hydrograph or reservoir storage equation.

There are many more numerical schemes available for ODEs, such as explicit, implicit, Adams–Bashford, or Adams–Moulton (Ascher & Petzold, 1998). A selection of the most used numerical solutions for ODE, as they are used in water related problems, is presented in this section of the book.

4.4.2 Explicit schemes (Euler method)

The problem defined by (4.26) can be solved using a discretization of the independent variable (i.e., time) between the values $t^{initial}$ and t^{final}. The discretization is made into N intervals of width Δt (see Figure 4.7).

$$t^n = t^{initial} + n \cdot \Delta t \tag{4.27}$$

where $\Delta t = (t^{final} - t^{initial})/N$; $n = 0, 1, \ldots, N$; and $u(t^n) = u^n$ is the value of $u(t)$ at the n-th time-step (see Figure 4.7).

In order to approximate the derivative du/dt, the *forward difference* expression for the derivative

$$\frac{du}{dt} = \frac{u^{n+1} - u^n}{\Delta t} + O(\Delta t) \tag{4.28}$$

is substituted into equation (4.26). It yields:

$$\frac{u^{n+1} - u^n}{\Delta t} + O(\Delta t) = f(u^n, t^n) \tag{4.29}$$

Re-arranging expression (4.29) gives:

$$u^{n+1} \approx u^n + \Delta t \cdot f(u^n, t^n) \tag{4.30}$$

It is important to note that $O(\Delta t)$ term has been neglected and by that the equation is now an *approximation*. Equation (4.30) provides an expression to calculate $u(t)$ at a particular time-step t^{n+1} from its value at a previous time-step, t^n. This is known as *explicit differencing numerical scheme*.

If $u^0 = u^{initial}$ is known than u^1 can be estimated; from u^1, estimation of u^2 can be done, and so on. After N steps, $u^N = u^{final}$ is determined.

In each approximation of the derivative term $O(\Delta t)$ is neglected therefore there is a small error, that is accumulating in time, hence, the error in the calculation of u^1 is carried over in the calculation of u^2, and so on (Figure 4.8).

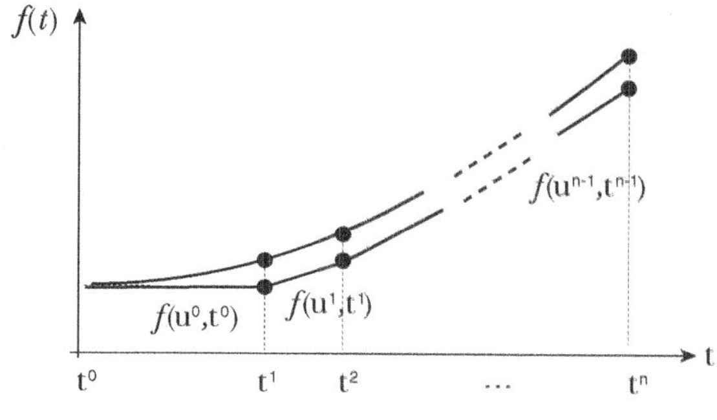

Figure 4.8 Error accumulation.

The step by step procedure of computing the values of u^n in time can be expressed as:

$$u^{n+1} \approx u^0 + \Delta t \cdot \sum_{j=0}^{n} f(u^j, t^j) \tag{4.31}$$

The explicit method approach for solving the ODE (4.26) is a one-step method approach.

4.4.3 Implicit schemes (Improved Euler method)

Using the same notations as in the previous section, in order to derive the implicit scheme, the term *du/dt* term in equation (4.26) is replaced by a forward differenced approximation and the value of the function $f(u, t)$ is computed at the $n + 1$ node of the discretization. This yields:

$$\frac{u^{n+1} - u^{n}}{\Delta t} + O(\Delta t) = f(u^{n+1}, t^{n+1}) \tag{4.32}$$

Re-arranging expression (4.32) gives:

$$u^{n+1} - \Delta t \cdot f(u^{n}, t^{n}) \approx u^{n} \tag{4.33}$$

This formula allows to calculate the unknown value of u^n at a certain time-step from its value at next time-step, u^{n+1}. Unfortunately to do this, value of u^{n+1} needs to be known and replaced into the function f in the first step of computation. This is what is known as an *implicit differencing scheme* and usually cannot be solved within a single step when the expression of the function f is complicated. Seldom it is necessary to use a technique called *iteration*, such as Newton's method.

4.4.4 Mixed schemes

If the explicit and explicit schemes are averaged together, the first-order ODE, defined by (4.26) is differenced as:

$$\frac{u^{n+1} - u^{n}}{\Delta t} + O(\Delta t) = \frac{1}{2}\left[f(u^{n}, t^{n}) + f(u^{n+1}, t^{n+1})\right] \tag{4.34}$$

By rearranging expression (4.34), it yields:

$$\frac{u^{n+1} - u^{n}}{\Delta t} \approx = \frac{1}{2}\left[f(u^{n}, t^{n}) + f(u^{n+1}, t^{n+1})\right] \tag{4.35}$$

Expression (4.35) is more complicated than the explicit or implicit schemes and may require iteration to be solved. It is known as *trapezoidal differencing scheme* or a *Crank-Nicholson* scheme.

4.4.5 Weighted averaged schemes

A more general formulation of the explicit, implicit and mixed schemes is the *weighted averaged* formulation, which gives one single formula for all three formulations. Consider as point of interest t^P, which is not positioned at a discretization point. Position of t^P inside the interval (t^n, t^{n+1}) is defined by the weighting parameter θ, which gives its position with respect to the nodes of the discretization grid (see Figure 4.9).

$$\theta = \frac{t^{P} - t^{n}}{\Delta t} \tag{4.36}$$

with θ ranging from 0 to 1.

Approximation of equation (4.26), using point P gives:

$$\frac{u^{n+1} - u^{n}}{\Delta t} \approx f(u^{P}, t^{P}) \tag{4.37}$$

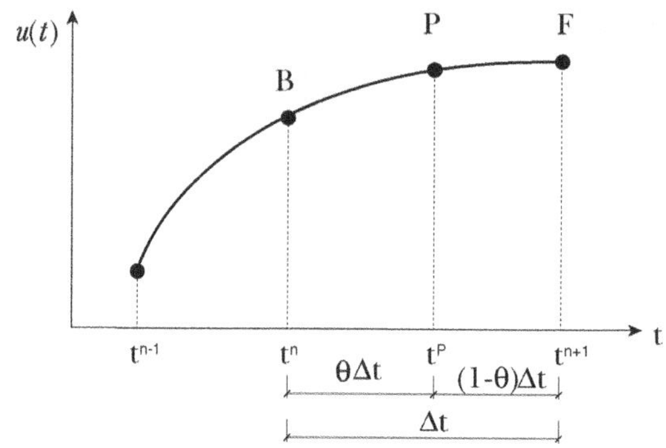

Figure 4.9 Weighted average position of a Point P in the interval (t^n, t^{n+1}).

The value of $f(u, t)$ in point P is calculated by linear interpolation between the end points of the interval as:

$$f(u^P, t^P) = (1-\theta)f(u^n, t^n) + \theta \cdot f(u^{n+1}, t^{n+1}) \tag{4.38}$$

which gives the final expression for u^{n+1} as:

$$u^{n+1} = u^n + \Delta t \cdot (1-\theta) \cdot f(u^n, t^n) + \Delta t \cdot \theta \cdot f(u^{n+1}, t^{n+1}) \tag{4.39}$$

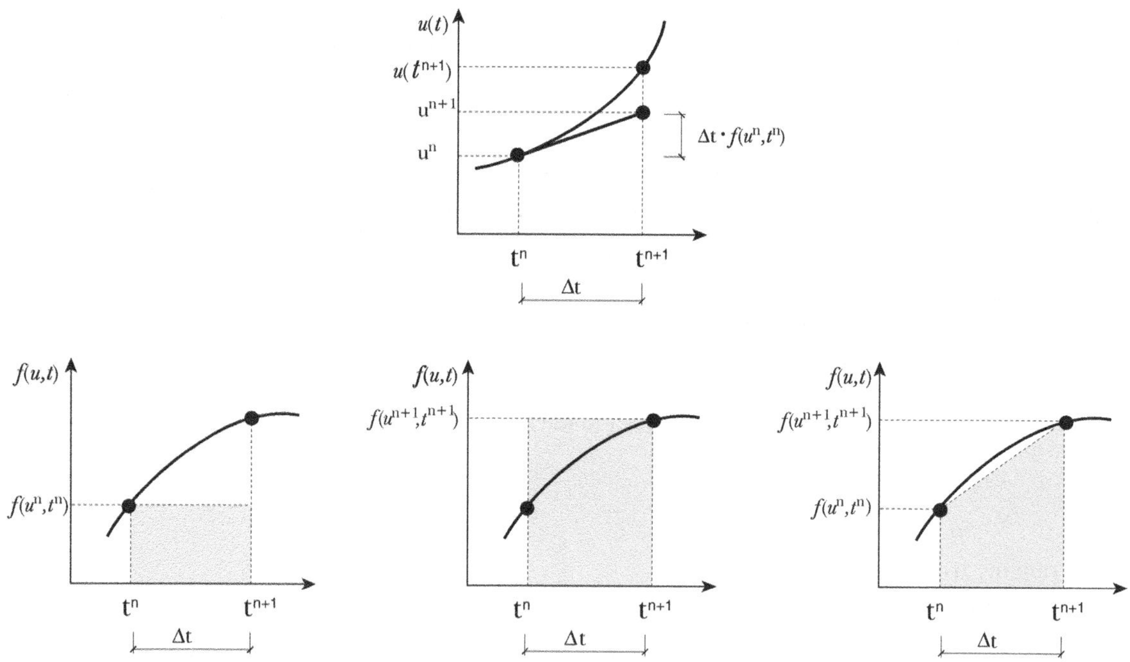

Figure 4.10 Integration of function $f(u, t)$ over interval Δt.

It can be noticed that scheme defined by expression (4.39) is:

- explicit Euler scheme, if $\theta = 0$;
- implicit scheme, if $\theta = 1$;
- mixed scheme, if $\theta = 0.5$;

Coefficient θ can have any value in the interval [0, 1], hence many more schemes can be defined. One of such scheme, known as Galerkin scheme is obtained for $\theta = 2/3$.

If the variation of function $f(u, t)$, from equation (4.26), is represented graphically, for a given $u(t)$, then it can be noticed that the difference between the three numerical schemes formulations, explicit, implicit and trapezoidal, is in the way the integration of $f(u, t)$ is done over the interval Δt. From Figure 4.10 it can be concluded that explicit approach is underestimating and implicit approach is overestimating the value of $f(u, t)$.

4.4.6 Runge–Kutta methods

A similar approach to the weighted averaged schemes are the so called Runge–Kutta methods in which a set of multiple internal points of computation are selected inside the interval (t^n, t^{n+1}).

The general numerical scheme for Runge–Kutta methods, for a number of m points inside the computational domain, can be formulated as:

$$\begin{cases} u^{n+1} = u^n + \Delta t \cdot \sum_{j=1}^{m} w_j k_j \\ k_1 = f\left(u^n, t^n\right) \\ k_j = f\left(u^n + \Delta t \cdot \sum_{l=1}^{j-1} a_{jl} k_l, t^n + c_j \cdot \Delta t\right) \end{cases} \tag{4.40}$$

where $m \geq 1$, and w_j, c_j, and a_{jl} are numerical coefficients. Depending of the number of internal points m, the corresponding Runge–Kutta method is called **m**-stage.

Legras (1971) presents formulas for Runge–Kutta coefficients, up to 8-stage. One of the most popular Runge–Kutta formula is the one in 4-stages for which the formulas defined by equations (4.40) are:

$$\begin{cases} k_1 = f(u^n, t^n) \\ k_2 = f\left(u^n + k_1 \dfrac{\Delta t}{2}, t^n + \dfrac{\Delta t}{2}\right) \\ k_3 = f\left(u^n + k_2 \dfrac{\Delta t}{2}, t^n + \dfrac{\Delta t}{2}\right) \\ k_4 = f\left(u^n + k_2 \Delta t, t^n + \Delta t\right) \end{cases} \tag{4.41}$$

with the solution at time step $n + 1$:

$$u^{n+1} = u^n + \frac{\Delta t}{6}\left(k_1 + 2k_2 + 2k_3 + k_4\right) \tag{4.42}$$

There are many more numerical schemes available for ODEs, such as Adams–Bashford or Adams–Moulton (Ascher & Petzold, 1998).

4.5 NUMERICAL SCHEMES FOR PARTIAL DIFFERENTIAL EQUATIONS

The principle of the finite difference methods is the same for PDEs as for ODEs, however PDEs are more complex to solve than the ODEs, because the unknown function that needs to be determined is differentiated with respect to both time and space. The boundaries, and the conditions which the solutions must satisfy on the boundaries, have considerable influence on the solution of a PDE. The influence of boundary conditions on the solution of PDEs is very important because a PDE might have solutions for a certain set of boundary conditions and might be insolvable for another set.

As it has been shown in Chapter 3, PDE's are of three kind. Each of them is be treated separatly in this section of the book. Due to the special importance of the hyperbolic and parabolic equations in fluid dynamics, emphasis is given to these two kind of equations. Principle of the FDM approach is demonstrated on 1D types of equations, which is the same for 2D and 3D. In case of PDEs, 1D refers strictly to the space component, time independent variable is the second independent variable with respect to which the unknown function is defined.

4.5.1 Principle of FDM for PDEs

The solution of a PDE cannot be determined analytically, if the PDE is defined on a finite domain. Numerical methods determine a solution of a PDE by replacing the unknown function continuous derivatives with discrete approximations. The solution is found at a number of discretization grid points in space and time (see Figure 4.6). The numerical solution of a function $u(x,t)$, varying in one dimensional space and in time, is determined at grid points $x_i = i\Delta x$, and time points $n\Delta t$. The notation mentioned in equations (4.13)–(4.14) is used throughout the chapter, that is, $u_i^n = u(i\Delta x, n\Delta t) = u(x_i, t^n)$.

Any approximation that is used, in order to determine a solution for the engineering problems should be a good one hence it must be consistent, stable, and convergent to the analytical solution. This issue is detailed in Chapter 6 of this book.

The principle of the FDM for PDEs is demonstrated considering the simple PDE of the scalar *linear advection equation,* which is of hyperbolic type:

$$\frac{\partial u}{\partial t} + a\frac{\partial u}{\partial x} = 0 \tag{4.43}$$

where a is the advection speed, assumed constant and positive. Appropriate boundary conditions are assumed to be specified at the upstream boundary. The unknown function $u(x, t)$ is one dimensional in space. The equation is linear because it consists of a sum of separate terms involving the advected quantity, u.

When the advection speed a is a constant, it can be demonstrated that any initial value of u is advected unchanged along the x axis. The general solution of the equation (4.43) is written as:

$$u(x,t) = F(x - at) \tag{4.44}$$

where $F(s)$ is an arbitrary single-valued differentiable function; with $s = x - at$. Substitution of expression (4.44) into the left-hand-side of equation (4.43) yields:

$$\frac{\partial F(s)}{\partial t} + a\frac{\partial F(s)}{\partial x} = \frac{dF}{ds}\frac{\partial s}{\partial t} + a\frac{dF}{ds}\frac{\partial s}{\partial x} = \frac{dF}{ds}\left(\frac{\partial s}{\partial t} + a\frac{\partial s}{\partial x}\right) = \frac{dF}{ds}(-a + a + 1) = 0 \tag{4.45}$$

Deduction (4.45) shows that $u(x, t)$ is carried along x at a constant velocity a, and the shape of $u(x, t)$ remains unchanged (see Figure 4.11).

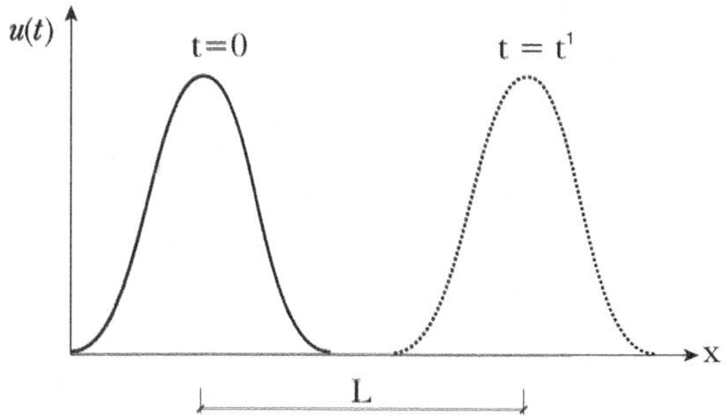

Figure 4.11 Advection of $u(x, 0)$ along the x axis.

In order to solve equation (4.43) using the discrete approach, the time-derivative between points (i, n) and $(i, n + 1)$ are used (see Figure 4.12):

$$\frac{\partial u}{\partial t} \cong \frac{u_i^{n+1} - u_i^n}{\Delta t^n} \tag{4.46}$$

along with the space-derivative:

$$\frac{\partial u}{\partial x} \cong \frac{u_i^n - u_{i-1}^n}{\Delta x_i} \tag{4.47}$$

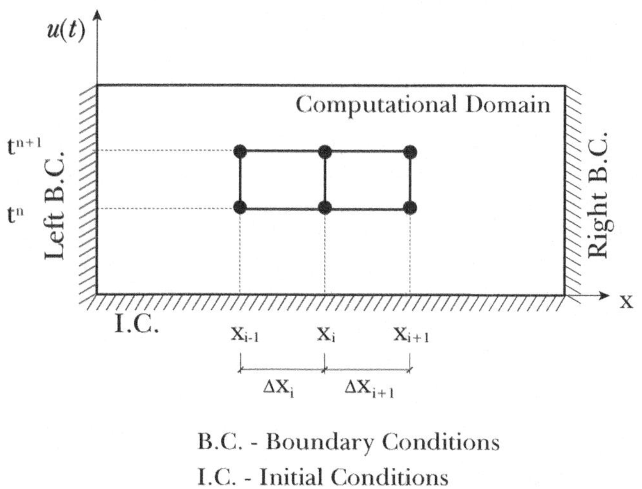

Figure 4.12 Unequally spaced grid discretization for the advection equation computational domain.

Replacing (4.46) and (4.47) into (4.43) yields:

$$u_i^{n+1} = Cr_i^n \cdot u_{i-1}^n + (1 - Cr_i^n) \cdot u_i^n \tag{4.48}$$

where:

$$Cr_i^n = a \cdot \frac{\Delta t^n}{\Delta x_i} \tag{4.49}$$

Expression (4.49) is called the *Courant number*, and depends on the grid size at time level $(n + 1)$ and space size at $(i + 1)$. Assuming that the values for u are known at any point i of the space, at time level n, the value for u at any point in space, at time level $n + 1$ can be computed using (4.48). Equation (4.48) cannot be applied at the left-hand boundary of the domain where $i = 1$, because there is no point to the left of the left boundary. This analysis shows the need to know the *Boundary Condition*, u_1^{n+1}, at the left boundary (see Figure 4.13).

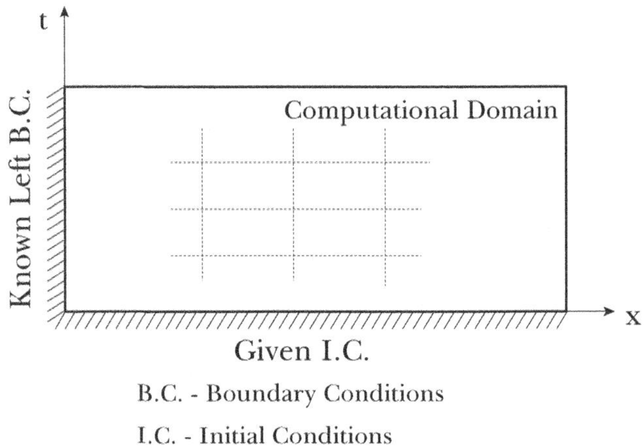

Figure 4.13 Definition of boundary conditions for advection-equation.

The notion of stencil is introduced here by definition. A stencil is the representation in the computational domain coordinates, of the nodes that relate to the point of computation by using a numerical approximation method. In case that two independent variables, space and time, are used to express a function in a differential equation, than in the notation used for stencils, $n - 1, n, n + 1$ indicate the time steps, and $i - 1$, i, $i + 1$ indicate space step. It is always assumed that all values of the function under consideration are known at time steps n and $n - 1$, and the time step $n + 1$ is to be calculated.

The stencil of the numerical scheme (4.48) is shown in Figure 4.14.

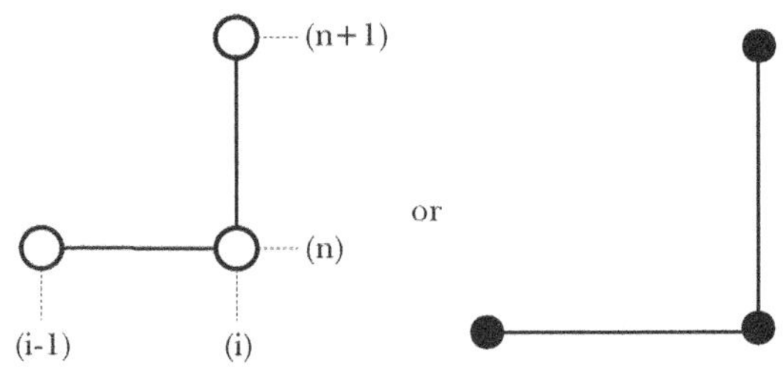

Figure 4.14 Example of a stencil for a numerical scheme.

4.5.2 Hyperbolic PDEs

Demonstration of the application of the FDM for hyperbolic equations is shown on the linear advection equation. As seen from the principle of the method the first step in applying the finite difference technique is to discretise the problem domain. Next step is to find a suitable approximation to the hyperbolic equation for the considered discretization.

The linear advection equation has two first order partial derivatives, $\partial u/\partial x$ and $\partial u/\partial t$. There are many different numerical schemes to determine the appropriate solution of the equation, depending on the selected approximation of the two derivatives, with respect to time and space. If only the three different approximations for such derivatives (forward, backward and central differences) are considered a minimum of nine different approximations to equation (4.43) can be expressed. The main question is which approximation should be used, because while seeking a solution one needs to be sure that the differencing scheme should be as accurate as possible, numerically stable and easy to implement.

4.5.2.1 Explicit and implicit schemes

As already mentioned in the beginning of the chapter, a scheme where the value at the unknown time level $(n + 1)$ can be computed directly from the previous time level is called an *explicit* scheme. Equation (4.48) demonstrated for the principle of the method for PDMs is an explicit type of scheme. Implicit schemes are further detailed bellow.

Given the domain of definition of the unknown variable $u(x, t)$, as defined in Figure 4.12, the first order space derivative of the unknown function is estimated using the values at time level $n + 1$, as:

$$\frac{\partial u}{\partial x} \cong \frac{u_i^{n+1} - u_{i-1}^{n+1}}{\Delta x_i} \tag{4.50}$$

Substituting expression (4.50) and (4.46) into (4.43) yields:

$$-Cr_{i-1}^{n+1}u_{i-1}^{n+1} + (1 + Cr_i^{n+1}) \cdot u_i^{n+1} = u_i^n \tag{4.51}$$

The stencil of the numerical scheme (4.51) is given in Figure 4.15.

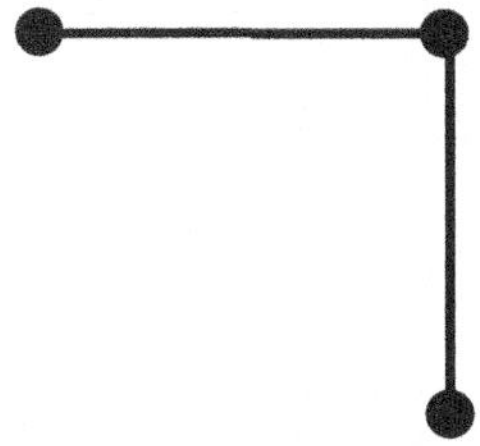

Figure 4.15 Stencil of the explicit scheme for advection equation.

The unknown solution at point i for time level $n + 1$ depends on the known values at the previous time level n, as well as on the unknown value at point $i - 1$ at time level $n + 1$. The values of u at time level $n + 1$ are linked together through a linear relationship. If equation (4.51) is written for all the nodes at time level $n + 1$, a system of equations is obtained and must be solved. The upstream boundary condition,

u_1^{n+1} is known and, u_2^n, is known from the previous time step, hence it is possible to compute, u_2^{n+1}. Sweeping along i in ascending order provides, u_3^{n+1}, u_4^{n+1}, and so on (see Figure 4.16).

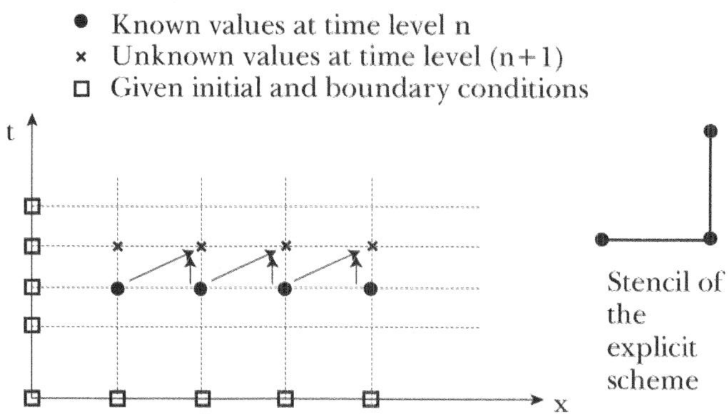

Figure 4.16 Explicit solution approach for hyperbolic equations.

A scheme where there is a relationship between several points at the *unknown* time level, leads to a system of equations. The equations of this system must be solved simultaneously to compute the unknowns. Such a scheme is called an *implicit* scheme. If the relationship between the unknown points is complex (for instance, non-linear), the system must be solved in an iterative way.

The differential approximations of the derivatives that are forward, backward and centred space are abbreviated as FS, BS and CS respectively. Similarly the ones for time are abbreviated as FT, BT and CT. Any numerical scheme that uses a combination of any of these two approximations is also referred to as a combination of the two abbreviations. The numerical scheme (4.49) is referred as FTBS, and (4.51) as BTBS.

4.5.2.2 *Forward time – centered space scheme (FTCS)*

The forward time, centred space scheme for the advection equation, based on the FT discretization of the time derivative of u, and CS discretization of the space derivative is:

$$\frac{u_i^{n+1} - u_i^n}{\Delta t} + a \cdot \frac{u_{i+1}^n - u_{i-1}^n}{2\Delta x} = 0 \tag{4.52}$$

or, re-arranging

$$u_i^{n+1} = u_i^n - \frac{1}{2} \cdot \frac{a\Delta t}{\Delta x}(u_{i+1}^n - u_{i-1}^n) \tag{4.53}$$

Durran (1999) shows through Fourier analysis that this scheme is unconditionally unstable. Numerical models still use a FTCS scheme to calculate the few first time-step, because the errors introduced by using the FTCS scheme at the beginning of the computational steps, are negligible when compared with all the other errors that are present in a model.

4.5.2.3 Centered and upwind schemes

The schemes define using the backward space approximation for the space derivative are called *upwind schemes* because the space information that is used to make the computations is coming from 'upstream' of the computational domain.

The schemes that are using centered derivatives, either for space or time are called *centered schemes*. One of the most interesting schemes of this kind is the CTCS scheme:

$$\frac{u_i^{n+1} - u_i^{n-1}}{2\Delta t} + a \cdot \frac{u_{i+1}^n - u_{i-1}^n}{2\Delta x} = 0 \tag{4.54}$$

or, re-arranging

$$u_i^{n+1} = u_i^{n-1} - \frac{a\Delta t}{\Delta x}(u_{i+1}^n - u_{i-1}^n) \tag{4.55}$$

Scheme defined by relation (4.55) can produce *two* solutions to the same problem under special circumstances of boundary conditions. The stencil of the CTCS scheme is given in Figure 4.17. It can be seen that for a particular grid-point i and time-step n, the CTCS approximation equation (4.55) links together the values of U at $(i, n + 1)$, $(i - 1, n)$, $(i + 1, n)$ and $(i, n - 1)$. The value of u at grid point (i, n) is not part of the relationship. The same equation for the same grid-point i at the next time-step, $n + 1$, links $(i, k + 2)$, $(i \pm 1, n + 1)$ and (i, n). Figure 4.17 shows the nodes for which the values of u effect the value of u at node (i, n) and those that effects the value of u at $(i, n + 1)$.

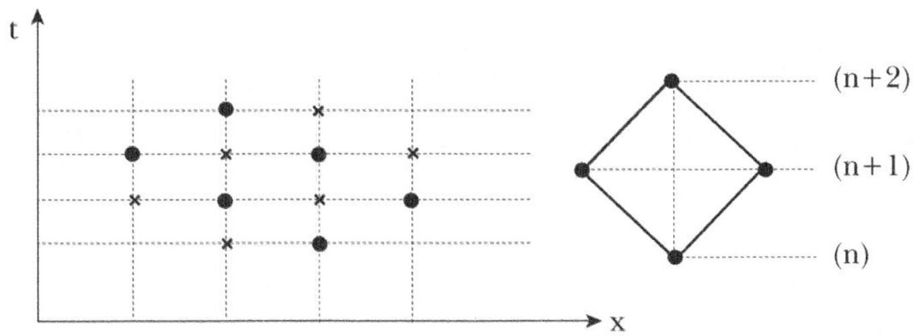

Figure 4.17 Relation between nodes for the CTCS numerical scheme.

4.5.2.4 The Preissmann scheme

In case of open channel flow modelling a frequently used numerical scheme is the Preissmann scheme, or 'box' scheme, or 4-point scheme. The scheme is implicit, very flexible and suitable for demonstrating some of the properties of the numerical solutions of hyperbolic types of equations (Abbott & Basco, 1989; Preismann, 1961). The scheme is developed based on four nodes of a discretization grid, as shown in (Figure 4.18). The derivatives are not approximated at the nodes of the grid, but at points located inside the grid.

Consider a grid equally spaced in space and time and the advection equation defined by (4.43). For a point P located between the points i and $(i + 1)$ in space and time levels n and $(n + 1)$ the following formulas are used to express the derivatives of an arbitrary function $f(x, t)$:

$$\left.\frac{\partial u}{\partial t}\right|_P = \psi \cdot \frac{u_{i-1}^{n+1} - u_{i-1}^n}{\Delta t} + (1 - \psi) \cdot \frac{u_i^{n+1} - u_i^n}{\Delta t} \tag{4.56a}$$

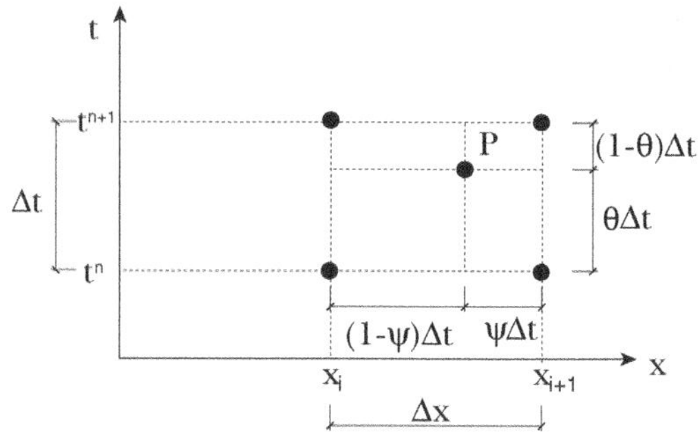

Figure 4.18 The four point (box) numerical scheme.

$$\left.\frac{\partial u}{\partial x}\right|_P = \theta \cdot \frac{u_i^{n+1} - u_{i-1}^{n+1}}{\Delta x} + (1-\theta) \cdot \frac{u_i^n - u_{i-1}^n}{\Delta x} \tag{4.56b}$$

where ψ and θ are weighting coefficients ranging from 0 to 1.

Formulas (4.56*a*, *b*) represents a weighted average of the derivatives between points $(i-1, n)$, (i, n), $(i, n+1)$ and $(i-1, n+1)$. If equation (4.43) is valid for a flood wave travelling in an open channel, then $(1-\psi)$ is the distance between point P and the cross-section positioned at grid point $(i-1)$ (Cunge *et al.*, 1986). For $\theta = 0$, the scheme is explicit, and it becomes fully implicit for $\theta = 1$. This scheme is unstable for $\theta < 0.5$.

Substitution of relations (4.56*a*, *b*) into (4.43), yields:

$$\psi \cdot \frac{u_{i-1}^{n+1} - u_{i-1}^n}{\Delta t} + (1-\psi) \cdot \frac{u_i^{n+1} - u_i^n}{\Delta t} + a\left((1-\theta) \cdot \frac{u_i^n - u_{i-1}^n}{\Delta x} + \theta \cdot \frac{u_i^{n+1} - u_{i-1}^{n+1}}{\Delta x}\right) = 0 \tag{4.57}$$

If all values of $u(x, t)$ at time level n are known then equation (4.57) relates the values of the unknown function $u(x,t)$ at grid points $(i, n+1)$ and $(i-1, n+1)$ as follows:

$$u_i^{n+1} = a_i \cdot u_{i-1}^n + b_i \cdot u_i^n + c_i \cdot u_{i-1}^{n+1} \tag{4.58}$$

with coefficients a_i, b_i, c_i, given by:

$$\begin{cases} a_i = \dfrac{\psi + (1-\theta) \cdot Cr}{1 - \psi + \theta \cdot Cr} \\ b_i = \dfrac{1 - \psi - (1-\theta) \cdot Cr}{1 - \psi + \theta \cdot Cr} \\ b_i = \dfrac{-\psi + \theta \cdot Cr}{1 - \psi + \theta \cdot Cr} \\ Cr = a\dfrac{\Delta t}{\Delta x} \end{cases} \tag{4.59}$$

In case that a canal of length L is considered, over a time period T, and an equally spaced grid discretization of the computational domain in $(M-1)$ space intervals Δx, and $(N-1)$ time intervals Δt.

Such a discretization results in M points along the x axis and N time levels of computation, as represented in Figure 4.19. Equation (4.57) is valid on every point $i = 2,\ldots, M$, at a specific time level n. At point $i = 1$, the value of function $u(x, t)$ is the upstream boundary condition and it is known, as are known all the values of $u(x, 0)$ as initial condition to the problem.

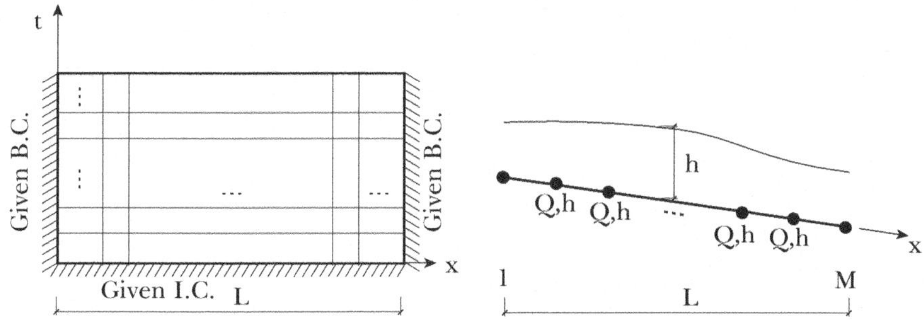

Figure 4.19 Computational domain for Preissmann scheme.

Writing equation (4.58) on every point $i = 2,\ldots, M$, a system of $(M - 1)$ equations, at time t^{n+1} is obtained. Adding the known boundary condition u_1^{n+1}, a linear system of M equations with M unknowns is required to be solved in order to have the solution.

4.5.2.5 The Abbott-Ionescu scheme for free-surface flow

The second scheme that is often used for free surface flows is the Abbott-Ionescu scheme, which was developed to solve the system of one-dimensional non-conservative shallow water equations (Abbott, 1979):

$$h\frac{\partial u}{\partial t} - u\frac{\partial h}{\partial t} + (gh - u^2)\frac{\partial h}{\partial x} = 0 \tag{4.60a}$$

$$g\frac{\partial h}{\partial t} - u\frac{\partial u}{\partial t} + (gh - u^2)\frac{\partial u}{\partial x} = 0 \tag{4.60b}$$

where g is gravitational acceleration; h channel water depth; and u flow velocity.

The discretization grid used for solving these equations is a staggered one that is, the space- and time-derivatives of the variables h and u are not computed at the same grid point (see Figure 4.20).

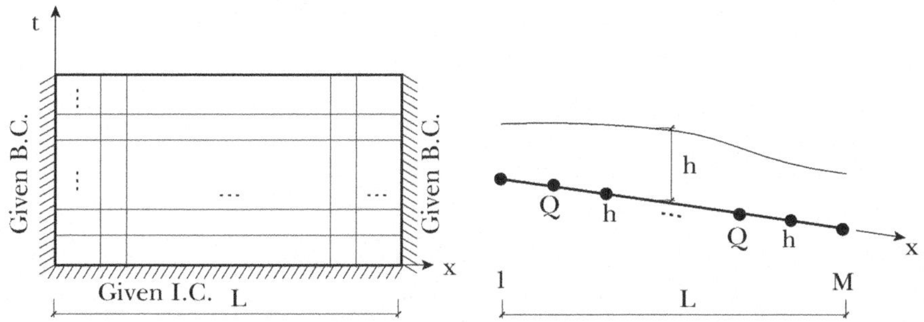

Figure 4.20 Computational domain for Abbott-Ionescu scheme.

The derivatives of the equation (4.60*a*) are defined as:

$$\frac{\partial h}{\partial t} = \frac{h_i^{n+1} - h_i^n}{\Delta t} \tag{4.61a}$$

$$\frac{\partial u}{\partial t} = \frac{1}{2} \cdot \left(\frac{u_{i-1}^{n+1} - u_{i-1}^n}{\Delta t} + \frac{u_{i+1}^{n+1} - u_{i+1}^n}{\Delta t} \right)0 \tag{4.61b}$$

$$\frac{\partial u}{\partial x} = \frac{1}{2} \cdot \left(\frac{u_{i+1}^{n+1} - u_{i-1}^{n+1}}{2\Delta x} + \frac{u_{i+1}^n - u_{i-1}^n}{2\Delta x} \right)0 \tag{4.61c}$$

The derivatives of the equation (4.60*b*) are defined as:

$$\frac{\partial u}{\partial t} = \frac{u_{i+1}^{n+1} - u_{i+1}^n}{\Delta t} \tag{4.61d}$$

$$\frac{\partial h}{\partial t} = \frac{1}{2} \cdot \left(\frac{h_i^{n+1} - h_i^n}{\Delta t} + \frac{h_{i+2}^{n+1} - h_{i+2}^n}{\Delta t} \right)0 \tag{4.61e}$$

$$\frac{\partial h}{\partial x} = \frac{1}{2} \cdot \left(\frac{h_{i+2}^{n+1} - h_i^{n+1}}{2\Delta x} + \frac{h_{i+2}^n - h_i^n}{2\Delta x} \right)0 \tag{4.61f}$$

Figure 4.21 illustrates the principle of the Abbott-Ionescu scheme.

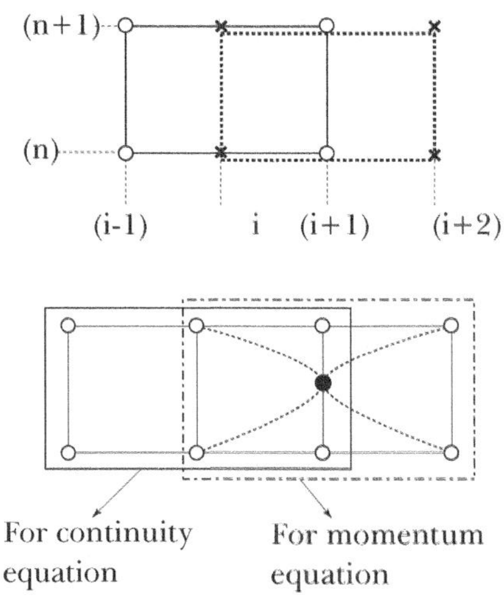

Figure 4.21 The Abbott-Ionescu numerical scheme.

4.5.3 Parabolic PDEs

The solution of parabolic equations using FDM approach is demonstrated on the one-dimensional diffusion equation, which has the general form:

$$\frac{\partial u}{\partial t} = D\frac{\partial^2 u}{\partial x^2} \tag{4.62}$$

where D is referred as the diffusion coefficient.

The same principle, as in case of hyperbolic equations apply, that is, a solution for the unknown function $u(x, t)$ is determined at time level $(n + 1)$, based on the values previously determined at time levels $1, 2, \ldots, n$.

In the practice of solving parabolic equations two types of boundary conditions are common to be defined for the domain of computation; Dirichlet boundary conditions; and Neumann boundary conditions (Fletcher, 1991). The Dirichlet boundary conditions are when the unknown function is defined at both ends of the computational domain, while Neumann boundary conditions are when the spatial derivative of the unknown variable is known at the both ends of the computational domain. An initial condition is also required in order to find the solution of the diffusion equation (Figure 4.22).

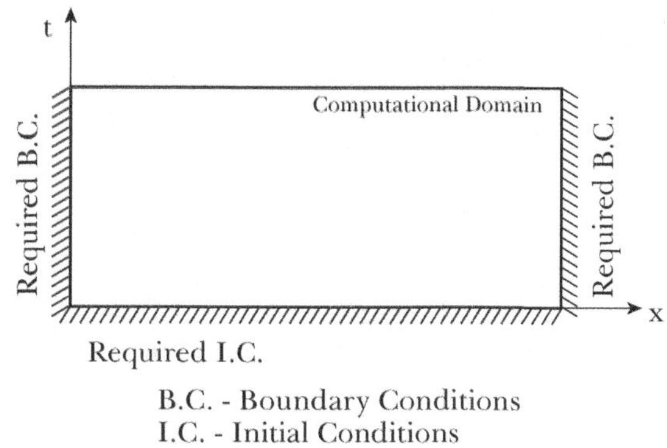

Figure 4.22 Computational domain, boundary and initial conditions for the diffusion equation.

4.5.3.1 *Explicit schemes*

The parabolic equation (4.62) involves a second-order space derivative. According to relation (4.23) at computational time level $(n + 1)$, the values of the unknown function in three points in space are needed to estimate the second-order space derivative. This is a CS approximation.

Consider an equally spaced computational domain (see Figure 4.23), a forward time approximation for the first order time derivative of $u(x, t)$; and a central space approximation for the space derivative (equation (4.23)). Replacement of the approximate derivatives in (4.62) yields:

$$\frac{u_i^{n+1} - u_i^n}{\Delta t} - D\frac{u_{i+1}^n - 2u_i^n + 2u_{i-1}^n}{(\Delta x)^2} = 0 \tag{4.63}$$

Re-arranging gives:

$$u_{i+1}^n = r \cdot u_{i-1}^n + (1 - 2r) \cdot u_i^n + r \cdot u_{i+1}^n \tag{4.64}$$

with

$$r = D \cdot \frac{\Delta t}{(\Delta x)^2} \tag{4.65}$$

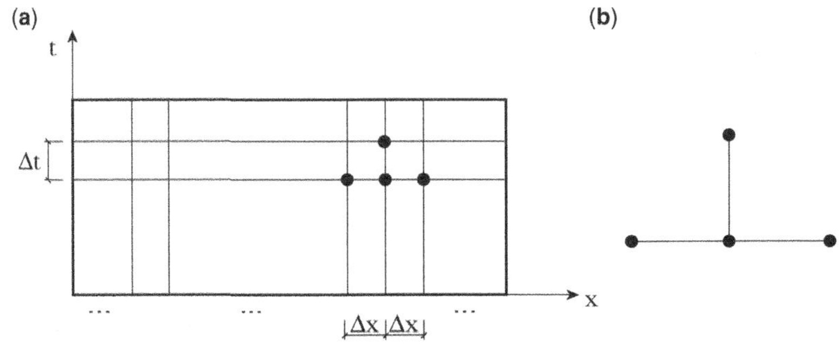

Figure 4.23 Stencil of the explicit FTCS scheme for a parabolic equation.

The stencil of the numerical scheme given by equation (4.64) is shown in Figure 4.24.

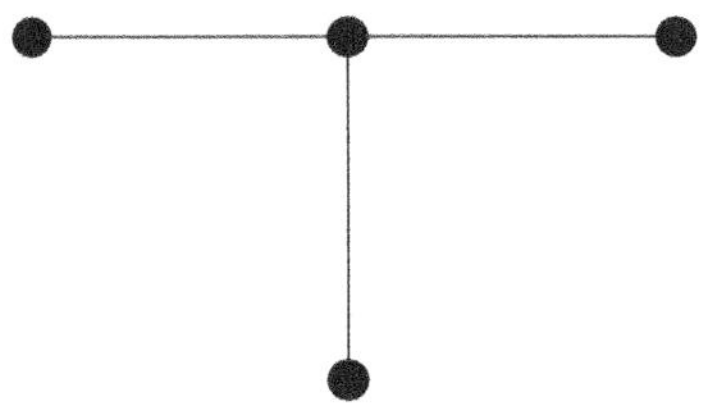

Figure 4.24 Stencil of the implicit scheme for a parabolic equation.

Other explicit schemes for the diffusion equation are the DuFort-Frankel or Three level scheme for which the corresponding equations are (4.66) and (4.67), respectively:

$$u_i^{n+1} = \left(\frac{2r}{1+2r}\right)\left(u_{i-1}^n + u_{i+1}^n\right) + \left(\frac{1-2r}{1+2r}\right)u_i^{n-1} \tag{4.66}$$

$$u_i^{n+1} = \left(\frac{1+2\gamma}{1+\gamma}\right)u_i^n - \left(\frac{\gamma}{1+\gamma}\right)u_i^{n-1} + \left(\frac{r}{1+r}\right)L_{xx}((1-\beta)u_i^n + \beta u_i^{n-1}) \tag{4.67}$$

with coefficients $\gamma \beta L_{xx}$ given by:

$$\beta = -0.5 - \gamma - \frac{1}{12r} \tag{4.68}$$

$$L_{xx} \cdot u_i^n = u_{i-1}^n - 2u_i^n + u_{i+1}^n \tag{4.69}$$

The value of the coefficient γ is arbitrary and has to be chosen with care so that in combination with the value of r fives a stable numerical scheme.

Equations (4.66) and (4.67) involves values of the unknown function at three time levels $(n - 1)$, n, and $(n + 1)$, except for the case $r = 0.5$ when equation (4.66) reduces to the FTCS scheme.

4.5.3.2 Implicit scheme

The implicit scheme for diffusion equation is obtained applying the FT derivative at time level n, in point i of the space, and the CS discretization of the derivative applied at time level $(n + 1)$ in point i of the space. The obtained FTCS implicit numerical scheme is:

$$r \cdot u_{i-1}^{n+1} + (1 + 2r) \cdot u_i^{n+1} - r \cdot u_{i+1}^{n+1} = u_i^n \tag{4.70}$$

The stencil of the relation (4.70) is shown in Figure 4.24.

The substitution of the derivatives in equation (4.62) leads to a system of equations, because numerical scheme (4.70) has to be written in all i nodes of the discretization grid, at time level $(n + 1)$. The resulting system of equation, that provides the final solution is:

$$\begin{bmatrix} (1+2r) & -r & \cdots & \cdots \\ -r & (1+2r) & -r & \cdots \\ \cdots & \cdots & \cdots & \cdots \\ \cdots & \cdots & \cdots & \cdots \\ \cdots & -r & (1+2r) & -r \end{bmatrix} \cdot \begin{bmatrix} u_1^{n+1} \\ u_2^{n+1} \\ \cdots \\ \cdots \\ u_M^{n+1} \end{bmatrix} = \begin{bmatrix} u_1^n \\ u_2^n \\ \cdots \\ \cdots \\ u_M^n \end{bmatrix} \tag{4.71}$$

Similar to the case of explicit schemes, a more general formulation of the implicit scheme is the three level scheme:

$$\frac{(1+\gamma) \cdot (u_i^{n+1} - u_i^n) - \gamma \cdot (u_i^n - u_{i-1}^n)}{\Delta t} - \alpha \cdot [(1-\beta)L_{xx} \cdot u_i^n + \beta L_{xx} \cdot u_i^{n+1}] = 0 \tag{4.72}$$

A particular Three level scheme, the one for which $\gamma = 0.5$ and $\beta = 1$, is often used for solution of problems.

4.5.4 Elliptic PDEs

As for previous types of PDEs, the application of the FDM method for elliptic type of PDEs is shown on the example of Laplace equation. Poisson equation is an example of an elliptic equation, and it is written in the form:

$$\frac{\partial^2 u}{\partial y^2} + \frac{\partial^2 u}{\partial x^2} = f(u, x, y) \tag{4.73}$$

In case that the function $f(u, x, y) = 0$ the equation is called Laplace.

Notice that time is not involved in the elliptic type of equations, because they are for steady state problems, hence time independent. For the numerical solution the domain represented in Figure 4.25 is used and the partial derivatives are substituted by the finite difference approximation (4.23). Because time is not involved in this kind of problem, only subscript notations are used, the ones for space (i and j).

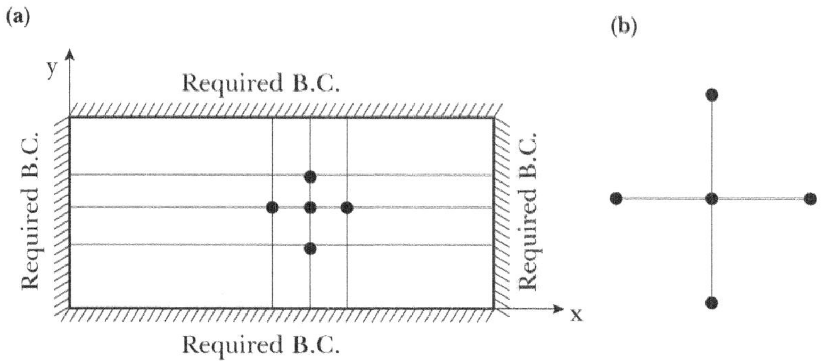

Figure 4.25 Domain of computation and stencil for elliptic type of equations.

For an equally spaced grid, (i.e., $\Delta x = \Delta y$) ,the numerical scheme obtained after substitution is:

$$u_{i+1,j} + u_{i+1,j} + u_{i+1,j} + u_{i+1,j} - 4u_{i+1,j} = 0 \tag{4.74}$$

The stencil corresponding to numerical scheme (4.74) is represented in Figure 4.25b.

Similar to the implicit methods for parabolic equations relationship (4.74) needs to be written for all interior points of the domain and the result is a system of equations that gives the solution for the unknown function. Relation (4.74) is often referred to as the *Laplacian difference equation.*

Analysing relation (4.72), it can be noticed that boundary condition along all the four edges of the computational domain must be specified in order to obtain a unique solution of the system of equation.

Boundary conditions at which derivatives of the unknown function are specified makes the problem solving more complex.

4.6 EXAMPLES

4.6.1 ODE: Solution of the linear reservoir problem

One of the most used ODE in hydraulic engineering is the linear reservoir. The linear reservoir is the continuity equation written for an inflow and outflow that changes the volume of water contained in a reservoir. The equation applies in reservoir operation, hydrological routing of flow, groundwater flow, and so on.

Given a reservoir, as shown in Figure 4.26, the change in volume is given by:

$$\frac{dV(t)}{dt} = Q_{in}(t) - Q_{out}(t) \tag{4.75}$$

in which $V(t)$ is the water volume stored in the reservoir, $Q_{in}(t)$ is the inflow into the reservoir and $Q_{out}(t)$ is the outflow from the reservoir.

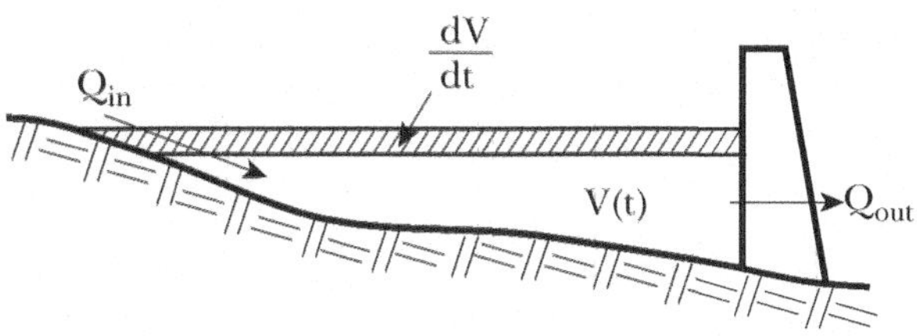

Figure 4.26 Scheme of a linear reservoir.

Equations (4.75) has two unknowns, the volume of the reservoir, $V(t)$, and the outflow from the reservoir $Q_{out}(t)$. Taking into account that the outflow from the reservoir is a function of the storage of the reservoir:

$$Q_{out}(t) = f(V(t)) \tag{4.76}$$

If the assumption is made that $f(V(t))$ is a power function then it can be written:

$$Q_{out}(t) = k \cdot [V(t)]^a \tag{4.77}$$

where k is a storage constant, which shows how fast the storage is emptied and is measured in units of 1/time; and a is a constant parameter. For a particular reservoir both a and b need to be determined from specific particularities of the reservoir.

In general it is considered that $a = 1$ is a good value (Henderson, 1966; Sturm, 2001), in which case equation (4.77) can be rearranged as:

$$V(t) = \frac{1}{k} Q_{out}(t) \tag{4.78}$$

If the simplified notation $Q_{out}(t) = Q$ and $Q_{in}(t) = I$ is used; and replacing equation (4.78) into the initial mass conservation equation (4.75) gives:

$$\frac{d\left(\dfrac{Q}{k}\right)}{dt} = I - Q \tag{4.79}$$

Re-arranging gives:

$$\frac{1}{k}\frac{dQ}{dt} = I - Q \tag{4.80}$$

or

$$\frac{dQ}{dt} + kQ = kI \tag{4.81}$$

Equation (4.81) can be integrating applying the integrating factor method, for solving ODE. The equation (4.81) is multiplied with the term e^{tk}, hence:

$$e^{tk} \cdot \frac{dQ}{dt} + ke^{tk}Q = ke^{tk}I \tag{4.82}$$

The right-hand-side of the equation (4.82) represents the derivative of the function Qe^{tk}, hence integrating the equation, yields:

$$e^{tk}Q = \int ke^{tk}I\,dt \tag{4.83}$$

The final expression for the outflow is:

$$Q = \frac{k\int e^{tk}I\,dt}{e^{tk}} \tag{4.84}$$

Expression (4.84) is not easy to be used in practice, because $I(t)$ is a function that has discrete values. When $I(t)$ is known the solution for $Q(t)$ is determined by using numerical integration of the term $e^{tk}I$.

In the particular case that $I(t) = 0$, equation (4.81) reduces to a simple ODE, that can be numerically solved using finite difference methods. Assuming that $Q(t = 0) = Q_o$, the equation that is solved numerically is (see Figure 4.27):

$$\frac{dQ}{dt} = -kQ \tag{4.85}$$

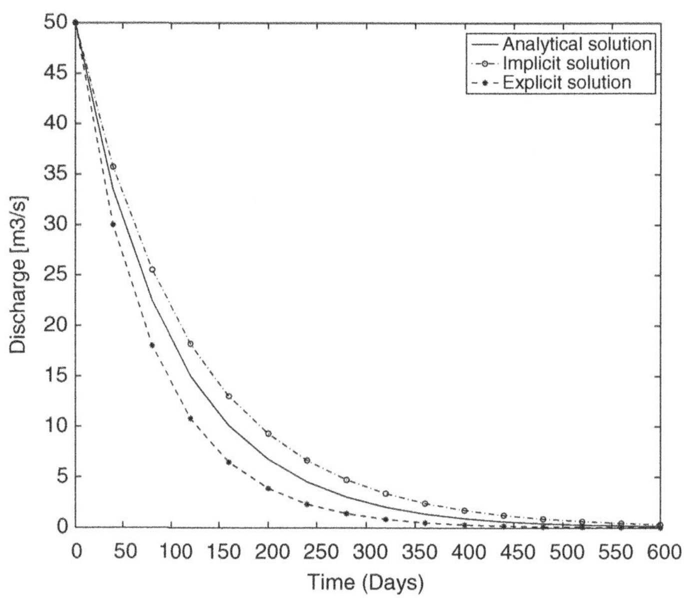

Figure 4.27 The analytical and numerical solution of the linear reservoir equation.

The explicit and implicit method are applied to solve the ODE defined by (4.85) and these are compared with the analytical solution in order to determine how well are the numerical approximations performing.

Application of an explicit numerical scheme to equation (4.85) gives:

$$\frac{Q^{n+1} - Q^n}{\Delta t} = -kQ^n \Rightarrow Q^{n+1} = Q^n(1 - k\Delta t) \tag{4.86}$$

Application of an implicit numerical scheme to equation (4.85) gives:

$$\frac{Q^{n+1} - Q^n}{\Delta t} = -kQ^{n+1} \Rightarrow Q^{n+1} = \frac{Q^n}{(1 + k\Delta t)} \tag{4.87}$$

Integration of equation (4.85) gives the analytical solution:

$$Q(t) = Q_o e^{tk} \tag{4.88}$$

Table 4.1 show the values of $Q(t)$ in case equations (4.83)–(4.85) are used to determine the solution. The values considered for Q_o, k, and Δt are:

$$Q_o = 50\frac{m^3}{s}; \quad k = 0.01\frac{1}{day}; \quad \Delta t = 40\,days \tag{4.89}$$

Figure 4.27 represents graphically the results presented in Table 4.1.

Table 4.1 Solution results for equation (4.85) for data as defined in (4.89) using analytical solution and explicit and implicit solution approaches.

Time (days)	Discharge (m³/s)		
	Analytical solution	Numerical solution	
		Explicit approach	Implicit approach
0	50.00	50.00	50.00
40	33.51	30.00	35.71
80	22.46	18.00	25.51
120	15.06	10.80	18.22
160	10.09	6.48	13.01
200	6.76	3.88	9.29
240	4.53	2.33	6.64
280	3.04	1.40	4.74
320	2.04	0.84	3.38
360	1.37	0.50	2.42
400	0.92	0.30	1.72

4.6.2 PDE: Simple wave propagation

A basic equation for the mathematical description of flood routing in a river is the conservation of mass equation:

$$\frac{\partial Q}{\partial x} + b\frac{\partial h}{\partial t} = 0 \tag{4.90}$$

where Q is the canal discharge; b the canal storage width; and h the water depth in the canal. Assuming that the relationship between Q and h is unique (i.e., unique stage – discharge relationship)-relation (4.90):

$$\frac{\partial Q}{\partial x} = \frac{dQ}{dh}\frac{\partial h}{\partial x} \tag{4.91}$$

equation (4.86) can be re-arranged as:

$$\frac{dQ}{dh}\frac{\partial h}{\partial x} + b\frac{\partial h}{\partial t} = 0 \tag{4.92}$$

or

$$\left[\frac{1}{b}\frac{dQ}{dh}\right]\frac{\partial h}{\partial x} + \frac{\partial h}{\partial t} = 0 \tag{4.93}$$

that simplifies to the linear advection equation:

$$\frac{\partial h}{\partial t} + c\frac{\partial h}{\partial x} = 0 \quad where \quad c = \frac{1}{b}\frac{dQ}{dh} \tag{4.94}$$

Using a discretization that is made up of a forward difference in time and backward difference in space, the solution of $h(x, t)$ for each grid point on the $x - t$ computational domain shown in Figure 4.28, which is 60 km long and has an input stage hydrograph raising from 2 m to 6 m over 6000 second. The total computational time is 60,000 seconds. Figure 4.29 and Table 4.2 shows the results obtained with the FTBS scheme for a grid size $\Delta x = 3000$ m and $\Delta t = 3000$ s; and a $c = 1$ m/s. Table 4.2 gives and extract for the first 50 minutes and for 24 km out of the 60 km long channel.

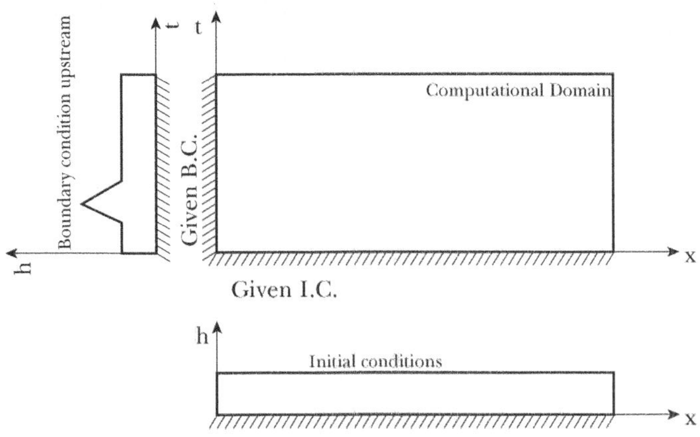

Figure 4.28 Computational domain, initial and boundary conditions for the simple wave propagation problem.

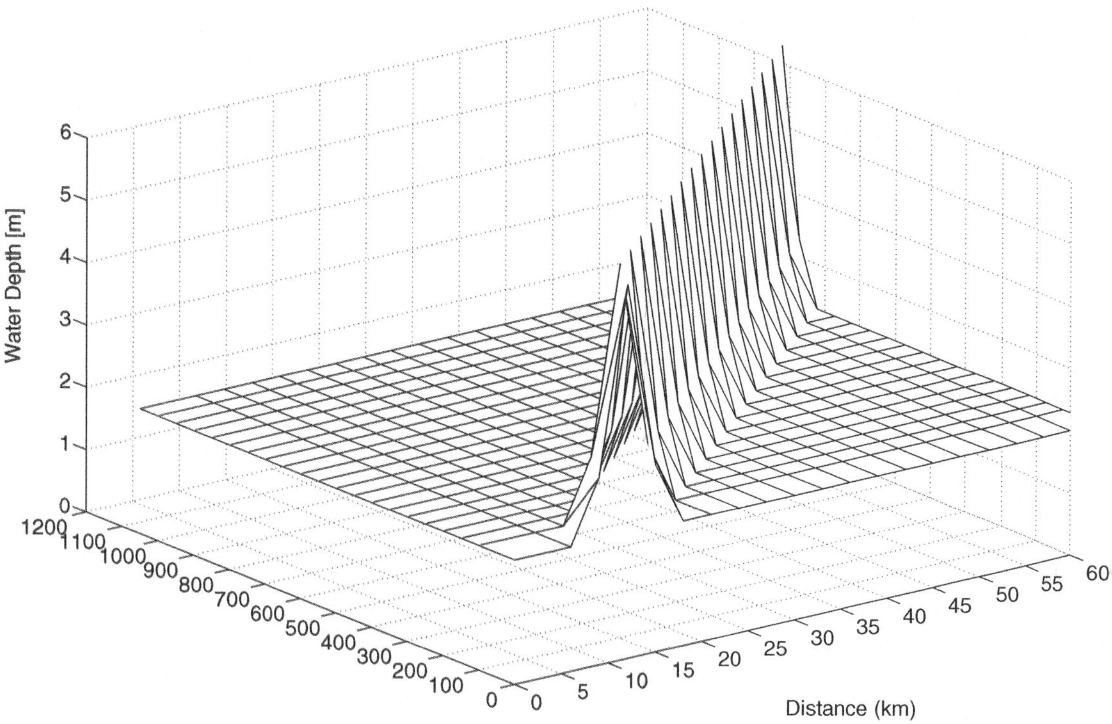

Figure 4.29 Solution of the simple wave propagation.

Table 4.2 Solution results (extract) for equation (4.94) for $\Delta x = 3000$ m and $\Delta t = 3000$ s, over 60 km long channel and during 60,000 seconds.

Time (min)	Water depth (m)								
	At Distance (km)								
	0	3	6	9	12	15	18	21	24
0	2	2	3	6	3	2	2	2	2
5	2	2	2	3	6	3	2	2	2
10	2	2	2	2	3	6	3	2	2
15	2	2	2	2	2	3	6	3	2
20	2	2	2	2	2	2	3	6	3
25	2	2	2	2	2	2	2	3	6
30	2	2	2	2	2	2	2	2	3
35	2	2	2	2	2	2	2	2	2
40	2	2	2	2	2	2	2	2	2
45	2	2	2	2	2	2	2	2	2
50	2	2	2	2	2	2	2	2	2

Results are showing that in case the product $c\ \Delta t/\Delta x$ is equal to one the advection equation transports the input unchanged. This product is called the CFL condition and it is described in detail in Chapter 6.

In case that this product is less than 1 diffusion appears in the solution, which is due to the numerical approximation and should not be confused with the attenuation of a flood wave. The result that is obtained for a selection of $\Delta x = 3000$ m and $\Delta t = 1500$, for the same canal, time length and $c = 1$ m/s is presented in Figure 4.30. The numerical effect of diffusion can be noticed.

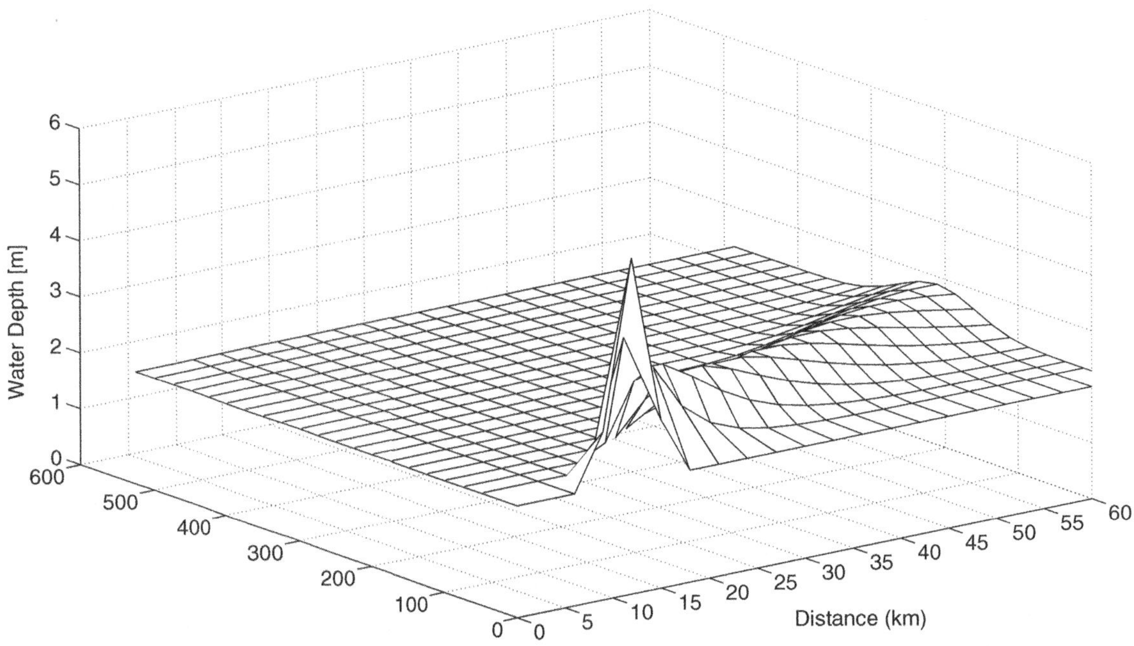

Figure 4.30 Solution of the simple wave propagation with numerical diffusion.

4.6.3 PDE: Diffusion equation

The 1D diffusion equation, of a pollutant, at a river section, where the pollutant is discharged is:

$$\frac{\partial C}{\partial t} = D_m \frac{\partial^2 C}{\partial x^2} \tag{4.95}$$

where C is chemical concentration, and D_m is the molecular diffusion coefficient. Equation (4.95) has several analytical solutions depending on initial and boundary conditions. For the case of a pollutant in the form of an impulse input at $x = 0$ into a semi-infinite media ($C = 0$ at $x = \infty$ at all times), where $C(t = 0) = 0$ at all other values of x, equation (4.95) has the following analytical solution:

$$C(x, t) = \frac{M}{2\sqrt{\pi \cdot D_m \cdot t}} e^{\frac{-x^2}{4 \cdot D_m t}} \tag{4.96}$$

where M is the mass of chemical added at $x = L$, with units of mass per cross-sectional area (e.g., mg/cm^2). This solution is valid when diffusion occurs in both directions, with the concentration profile extending further as time passes.

Using a discretization that is made up of a forward difference in time and backward difference in space, the solution of $c(x,t)$ for each grid point on the $x - t$ computational domain, is given Figure 4.31, for the grid size $\Delta x = 50$ cm, $\Delta t = 27$ hours; $D_m = 0.01$ cm^2/sec, $M = 0.7$ mg/cm^2. The length over which diffusion takes place is 600 cm.

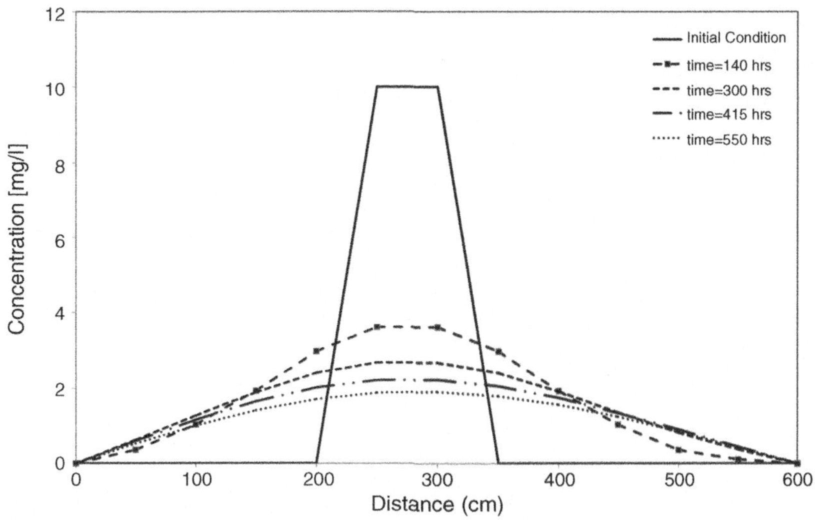

Figure 4.31 Initial condition and solution for diffusion equation.

Chapter 5

Finite volume method

5.1 GENERAL CONCEPT

The finite volume method (FVM) is a numerical approach applicable to differential equations that represent various types of conservation laws. As a method it was originally developed to solve water related problems and has been extended to other disciplines later. FVM has the advantage that it can be used on arbitrary geometries and on structured or unstructured discretization of the computational domain.

In the FDM approach, detailed in the previous chapter, the differential equation to be solved is written at each discretization node of the computational domain by replacing the derivatives with a finite difference approximation. The FDM method, however is difficult to apply when the coefficients involved in the differential equation are discontinuous (LeVeque, 2002). In the FVM method, discontinuities of the coefficients are not a problem if the computational grid is chosen in such a way that these appear at the boundaries of the computational domain (Durran, 1998). Finite difference approximations are sometimes used in FVM approach to approximate fluxes at the boundary of the *control volume*, the discretization unit of the computational domain. Hence the FVM differs from the FDM in the way finite difference approximation are used (i.e., for the flux of a quantity, rather than for a derivative).

Depending on how the discretization of the computational domain is done, the FVM is also known as cell centred, vertex centred, box method, or balance method.

Finite volume methods have been extensively developed by different researchers for each type of partial differential equation (hyperbolic, parabolic and elliptic) and dependent of the discretization type used; Lazarov and Makarov (1982) for Cartesian meshes, Mandel and McCormick (1989) for unstructured meshes; Manteuffel and White (1987) for cell-centred finite volume schemes; LeVeque (2002), Godlewski and Raviart (1996) for hyperbolic problems; and so on.

Finite volume methods can be characterised as follows:

- Problem domain is discretized into discrete elements called control volumes (CV);
- The differential equation to be solved is integrated into a balance equations for each control volume of the computational domain;
- Obtained integrals are approximated using numerical integration;

- The values of the unknown function and their derivatives are approximated using the values at the nodes of the control volume;
- The solution of the problem over the computational domain is obtained by assembling the obtained discrete equations over a control volume, into an algebraic system of equations.

An important feature of the FVM approach is the fact that the numerical flux is conserved from one discretization cell to another (the neighbouring cell), that is, there is local conservation of the numerical fluxes. This is very important especially because flux is an important element in fluid dynamics.

The FVM principle is detailed bellow. The method is applicable for the solution of conservation laws, and for simplicity of the demonstration of the method the one-dimensional general form of the conservation law (see chapter 2) is used:

$$\frac{\partial}{\partial t}\int_V u(x,\ t)dV + \int_A f(u)\cdot \boldsymbol{n}\cdot dA - \int_V S(u,\ t)dV = 0 \tag{5.1}$$

Equation (5.1) shows the transport of a variable (substance, water depth, etc.) under the influence of advection, that is, the rate of change of variable $u(x, t)$ inside a defined volume V together with the flux of $u(x, t)$ through boundary A is the same as the rate of $S(u, t)$. Equation (5.1) is valid at each point x and each time t, of the computational domain, where the conservation of $u(x, t)$ is written. $S(u, x, t)$ is also known as *source term* of the equation and expresses a possible volumetric exchange between conserved quantities $u(x, t)$, that is, the diffusion phenomena. The term $\boldsymbol{n}$ represents the unit vector perpendicular on surface A, pointing outwards.

Before showing the principle of the FVM method it is important to note that in the integral expression (5.1) the derivatives are zero order for the advection term ($f(u)$) and first order for the diffusion term ($S(x, t)$), if it exists. The fact that the order of the derivatives is so low is important when solutions are sought for problems where phenomena change rapidly in space so that the spatial derivative does not exist, such as in the case of a hydraulic jump. Functions that are discontinuous do not have derivatives at the discontinuity position, hence the conservation equation expressed as a PDE, is not valid at the discontinuity point. Finite volume method presents the advantage that they overcome this issues, by considering the integral representation of the PDE and by having reduced order of derivatives, that can be easily treated.

The principle of the method is illustrated on a simple example of the one-dimensional linear advection equation, which is obtained from (5.1) if $f(u) = au$ and $S(u, t) = 0$:

$$\int_V \left(\frac{\partial u}{\partial t} + \frac{\partial f(u)}{\partial x}\right) dV = 0 \tag{5.2}$$

Consider the above equation to be solved over the one dimensional computational domain V that is discretized in *control volumes* V_i, defined between points $x_{i-1/2}$ and $x_{i+1/2}$ (see Figure 5.1b). Volume integration of equation (5.2) over the control volume (CV) yields:

$$\int_{x_{i-1/2}}^{x_{i+1/2}} \frac{\partial u}{\partial t} dx + \int_{x_{i-1/2}}^{x_{i+1/2}} \frac{\partial f}{\partial x} dx = 0 \tag{5.3}$$

In Figure 5.1 points x_i are equally spaced along the space domain x with a constant distance $\Delta x = x_{i+1/2} - x_{i-1/2}$; $i = 1, \ldots, N-1$. Same convention of notation holds, as in Chapter 4, that $u(x_{i+1}) = u(x_i + \Delta x) = u_{i+1}$. In Figure 5.1 control volumes are noted with V_i.

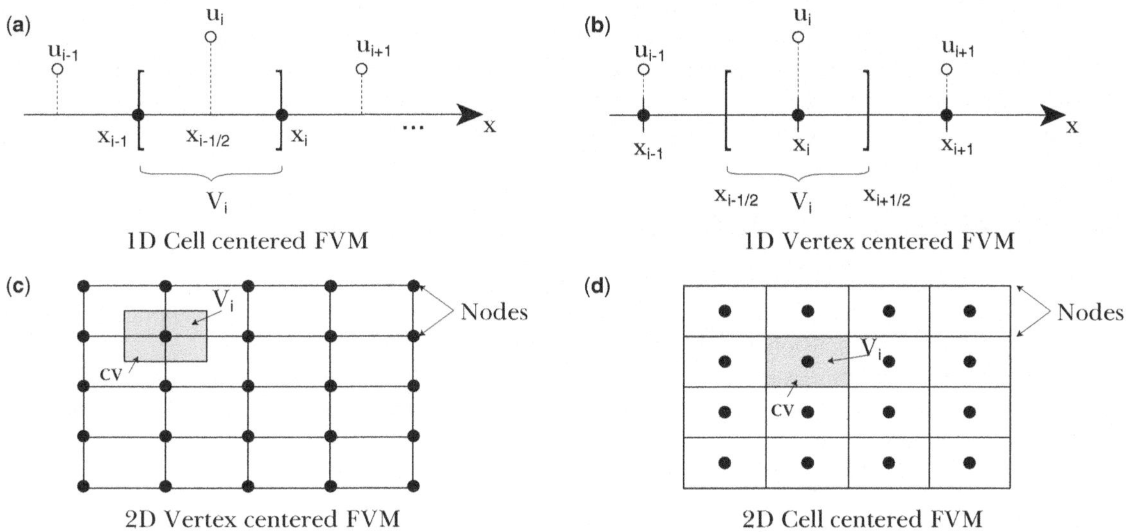

Figure 5.1 Finite volume examples of control volume in one- and two-dimensional space.

Equation (5.3) represents the integral formulation of the differential equation over control volume V_i which is written in all CVs of the computational domain to yield a system of equations, and gives the solution to the problem by determining the values of the unknown variable in all CVs.

Solution of the system formed by types of equation (5.3) requires:

- numerical approximation of the space integrals, which can be volume and/or surface integrals, depending on the problem dimension;
- interpolation, to determine the values of coefficients that describe the numerical approximations of the surface and volume integrals; and
- time integration by evaluating the time derivatives.

If u_i is averaged and considered constant over CV, then the first term of the equation (5.3) can be integrated numerically as:

$$\int_{x_{i-1/2}}^{x_{i+1/2}} \frac{\partial u}{\partial t} dx = u_i \cdot (x_{i+1/2} + x_{i-1/2}) \tag{5.4}$$

The second term of equation (5.3) integrated analytically yields:

$$\int_{x_{i-1/2}}^{x_{i+1/2}} \frac{\partial f}{\partial x} dx = f(u_{i+1/2}) - f(u_{i-1/2}) \tag{5.5}$$

Substitution of equation (5.4) and (5.5) into equation (5.3) gives its discrete form as:

$$u_i \cdot (x_{i+1/2} + x_{i-1/2}) + f(u_{i+1/2}) - f(u_{i-1/2}) = 0 \tag{5.6}$$

Equation (5.6) is a conservative scheme if the flux on the boundary of one cell is equal with the flux on the boundary of the neighboring cell.

The discretization of the space domain, as selected, is known as cell-centered method, where the unknown variables, $u(x, t)$ are computed as average values over the CV and they are attached to the middle of the cell. A similar discretization is the cell-vertex method, in which the variables are averaged over a CV, but they are attached to the nodes of the discretization grid, that is, at the edge of the CV (see Figure 5.1a). Figure 5.1c, d, shows two types of discretization for a two dimensional space.

5.2 FVM APPLICATION DETAILS

5.2.1 Step by step application of the FVM

Based on the FVM principle illustrated above, a general step by step procedure for the application of the method is given for a 3-dimensional problem.

Equations (5.1) and (5.2) are exact formulations of the problem of a phenomenon, that is, the conservation law principle of a quantity $u = u(x, y, x, t)$. (*Note*: for a 3-dimensional function $u(x, y, z, t)$ the notation $\boldsymbol{u}$, is used when reference is given to all 3 dimensions of u). Consider the 3-dimensional advection-diffusive equation, which is a parabolic type of differential equation:

$$\frac{\partial u}{\partial t} + \sum_{i=1}^{3} \frac{\partial}{\partial x_i}\left(u - \alpha \frac{\partial u}{\partial x_i}\right) = S(u) \tag{5.7}$$

where x_i are the (x, y, z) components of the space, t is time, α is the diffusion coefficient and $u = u(x, y, z, t)$ is the unknown transported quantity (see Figure 5.2).

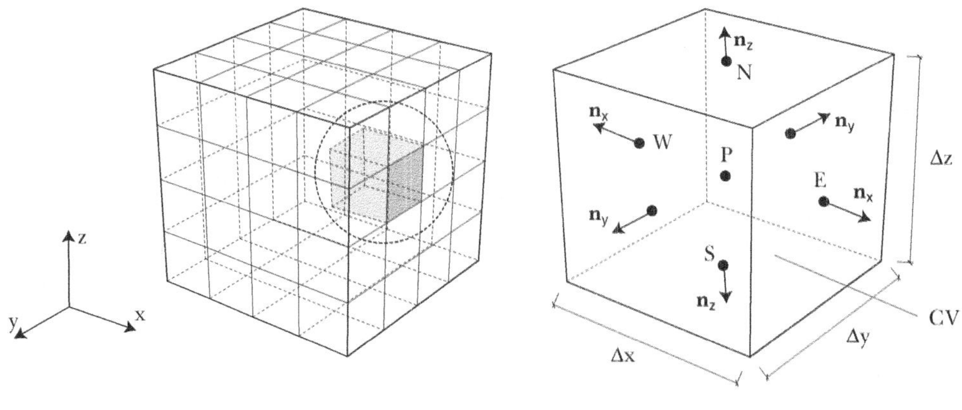

Figure 5.2 Computational domain for the 3D advection-diffusion equation.

The domain is divided into i $(i = 1, \ldots, N)$ computational cells (Vi) – control volumes. A number of nodes are created at the center of the control volumes, where the unknown variable is computed. The balance equation for each of the defined CV is obtained by formulating the problem in an integral form, which is equivalent to the integration of equation (5.7). The integration over an arbitrary CV yields:

$$\frac{\partial}{\partial t}\int_V u dV + \sum_{i=1}^{3}\int_A \left(u - \alpha \frac{\partial u}{\partial x_i}\right) n_i dA = \int_V S(u) dV \tag{5.8}$$

where, A is the surface of the CV and n_i the unit normal vector to the surface, corresponding to the x_i component of the space.

In the FVM approach a numerical approximation of the solution is introduced by the space integration of the equation, hence fluxes in space and time are to be calculated. At each CV the average value of the function $\boldsymbol{u}$ is known. The advection and diffusion fluxes of the equation (5.8) are calculated in three steps, as follows:

(1) Determination of the values of function $\boldsymbol{u}$ for each CV: Both the value of the function $\boldsymbol{u}$ and its derivatives are numerically approximated for each CV. Based on these approximations both advective and diffusive fluxes are determined. Function $\boldsymbol{u}_i$ for the CV i (Vi) is approximated using a polynomial representation:

$$u_i = \sum_{n=1}^{P} a_n \phi_n(x) \tag{5.9}$$

where a_n are multiplication coefficients of ϕ_n chosen interpolation functions (simple polynomials). Each ϕ_n is valid in one of the P surrounding cells of the considered Vi control volume. Coefficients a_n are determined from the condition that the cell averages are valid over a number of $(i + m)$ cells, hence:

$$\int_{V_{i+m}} u dV = V_{i+m} u_{i+m}, \quad m = 0, 1, 2, \ldots, P-1 \tag{5.10}$$

where V_{i+m} are the number of P cells surrounding the cell Vi.

The unknown coefficients, a_n, are determined from a system of algebraic equation, of the form:

$$\sum_{n=1}^{P} A_{m,n} a_n = V_{i+m} u_{i+m}, \text{ with } A_{m,n} = \int_{V_{i+m}} \phi_n dV \tag{5.11}$$

(2) Evaluation of the integrals: The integrals are approximated numerically. The most often used approach to evaluate the integrals is the Gauss quadrature approach. In this case the selected functions ϕ_n determines the number of quadrature points to match polynomials of degree P.

(3) Temporal integration: The computed fluxes are used to advance the solution in time using an explicit time approximation method such as Euler, or Runge–Kutta, or the Adams-Bashforth class of methods. Time integration is dependent on the spatial approximation and is checked for stability.

The details of the three calculation steps are further explained, along with considerations regarding boundary conditions and the solution of the obtained system of equations. The chapter ends with an example on how to apply the method for a one dimensional advection diffusion equation.

5.2.2 Surface and volume integrals

Consider the example of a two dimensional domain which is split in CV quadrilateral shapes. Figure 5.3 shows the form of such a CV and the notations used on it. The nodes defining the CV are SW, SE, NE, NW along with the midpoints of the sides; S, E, N, and W. A special point, P, is positioned at the center of the CV. Figure 5.3 includes the representation of the unit normal vectors.

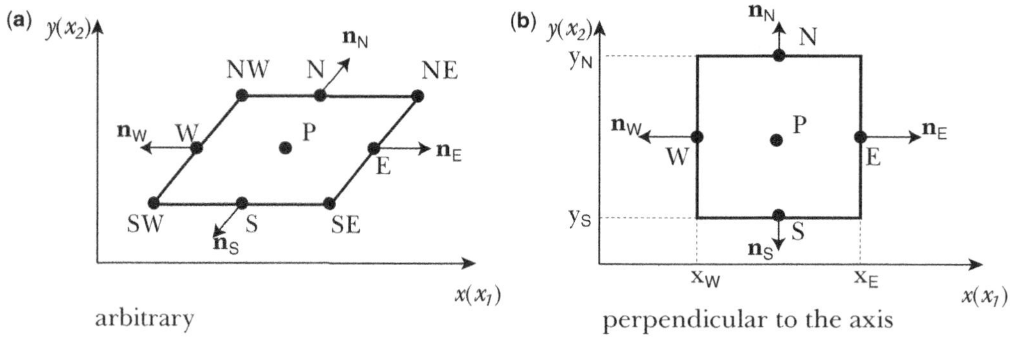

Figure 5.3 Quadrilateral control volume: (a) oblique; and (b) perpendicular on the coordinate axes.

According to the notation in Figure 5.3, the surface integral of equation (5.8) on the left-hand-side, is the summation of four surface integrals, over the four sides of the CV:

$$\sum_{l=1}^{4}\int_{A_l}\left(u-\alpha\frac{\partial u}{\partial x_i}\right)n_{li}dA_l \tag{5.12}$$

In equation (5.12) l is the number of sides of the CV (i.e., 4 sides), and all indices with l, represents the elements corresponding to side l of the CV. There are two types of fluxes; convective F_l^C, and diffusive F_l^D through the CV faces l:

$$F_l^C = \int_{A_l} un_{li}dA_l \tag{5.13}$$

and

$$F_l^D = -\int_{A_l}\left(\alpha\frac{\partial u}{\partial x_i}\right)n_{li}dA_l \tag{5.14}$$

The area/length of side A_l, is the length of the CV side. For the neighboring CVs with a common side the total value of the flux is:

$$F_l = F_l^C + F_l^D \tag{5.15}$$

The sign of the flux, however is opposite for the two neighboring cells.

The approximation of the surface integrals (5.13) and (5.14), for side l of the CV with a cell-centered variable, is done by steps 1 and 2, as detailed in Section 5.2.1. It yields:

$$\begin{cases} F_l^C = m_l\cdot\Delta A_l\cdot u, \\ F_l^D = \alpha\cdot\Delta A_l\cdot\left(\dfrac{\partial u}{\partial x_i}\right)_l \end{cases} \tag{5.16}$$

where m_i is the mass through face l of the CV. Relation (5.16) is approximated at side AE of the CV for a selected polynomial. ΔA_l is the area of element l.

The volume integral of equation (5.9) is approximated through numerical integration. The following relation holds over a CV:

$$\int_V S dV \approx S_P \Delta V \tag{5.17}$$

where S_P represents the value of S at the center of the CV (the average value of S over CV). ΔV represents the volume of the considered CV. Taking into account notations from Figure 5.3 the volume of the CV can be expressed as:

$$\Delta V = \frac{1}{2}\left|(x_{AE} - x_{NW})(y_{NE} - y_{Aw}) - (x_{NE} - x_{AW})(y_{AE} - y_{NW})\right| \tag{5.18}$$

All the integrals in equation (5.9) evaluated using (5.17) and (5.18) are added up over the whole computational domain. The following overall equation holds:

$$\underbrace{\sum_{cv} m_{cv} \Delta A_{cv} u_{cv}}_{convective\ fluxes} - \underbrace{\sum_{cv} \alpha \cdot n_{cvi} \Delta A_{cv} \left(\frac{\partial u}{\partial x_i}\right)_{cv}}_{diffuzive\ fluxes} = \underbrace{S_P \Delta V}_{source\ terms} \tag{5.19}$$

The values of the unknown function and its derivatives at each side of the CVs are determined using equation (5.19).

5.2.3 Discretization of convective fluxes

The convective fluxes, F^C are approximated using the values of function u at the CV center. The values of u at the CV center are calculated using approaches similar to FDM; upwind techniques, central differences, mixed techniques, and so on however they should be regarded as interpolation techniques, not finite differences. Interpolation is done in FVM usually in the direction of flow, that is, the assumption is made that the convective transport of u only takes place towards downstream, which is the case when the equation is of the parabolic type as is the case of the advection diffusion equation. Details on approximating the integrals are shown bellow on a one-dimensional equally spaced computational domain.

5.2.3.1 Upwind approach

The upwind method (UDS) approximates the values of u by a step function. The value of u over a CV on the E side of the CV (u_E) is determined as follows (see Figure 5.4):

$$u_E = u_P, \quad if\ m_E > 0 \tag{5.20a}$$

$$u_E = u_R, \quad if\ m_E < 0 \tag{5.20b}$$

A special upwind approximation is the quadratic upwind interpolation, also known as the QUICK method (see Figure 5.5). In the QUICK approach a polynomial is determined using an interpolation through 3 points; two neighboring points P and R, and a third point, RE or PW, depending on the flow direction. The resulting polynomial for the approximation of u_E is:

$$u_E = a_1 u_R - a_2 u_{PW} + (1 - a_1 + a_2) u_P, \quad if\ m_E > 0 \tag{5.21a}$$

$$u_E = b_1 u_P - b_2 u_{RE} + (1 - b_1 + b_2) u_R, \quad if\ m_E < 0 \tag{5.21b}$$

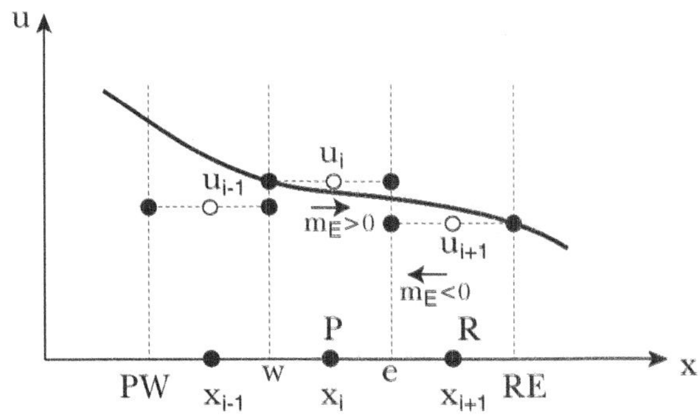

Figure 5.4 Constant approximations of u over the control volume.

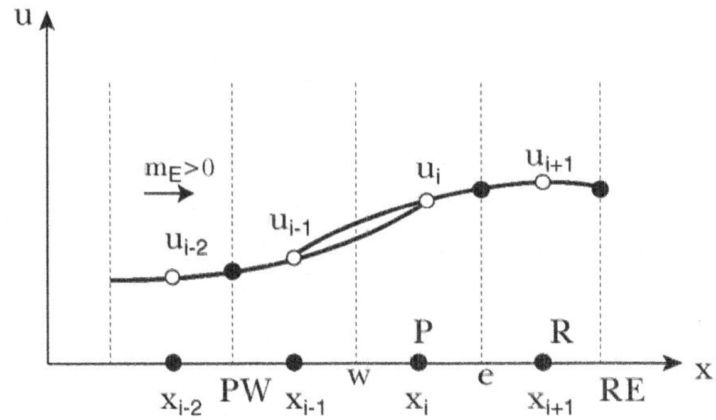

Figure 5.5 Approximation of function u using linear interpolation over two control volumes.

with values of the coefficients given by:

$$a_1 = \frac{(2-\gamma_w)\gamma_e^2}{1+\gamma_e-\gamma_w}, \qquad a_2 = \frac{(1-\gamma_e)(1-\gamma_w)^2}{1+\gamma_e-\gamma_w} \tag{5.22a}$$

$$b_1 = \frac{(1+\gamma_w)(1-\gamma_e)^2}{1+\gamma_{re}-\gamma_e}, \qquad b_2 = \frac{\gamma_{re}^2\gamma_e}{1+\gamma_{re}-\gamma_e} \tag{5.22b}$$

and,

$$\gamma_e = \frac{x_e - x_P}{x_R - x_P} \tag{5.23}$$

Relation (5.24) represents the interpolation factor for function u at side e, based on the values of u in points P and R (see Figure 5.6), that is:

$$u_e \approx \gamma_e u_e + (1 - \gamma_e)u_P \tag{5.24}$$

In case of a equally spaced grid: $a_1 = \frac{3}{8},\ a_2 = \frac{1}{8},\ b_1 = \frac{3}{8},\ b_2 = \frac{1}{8}.$

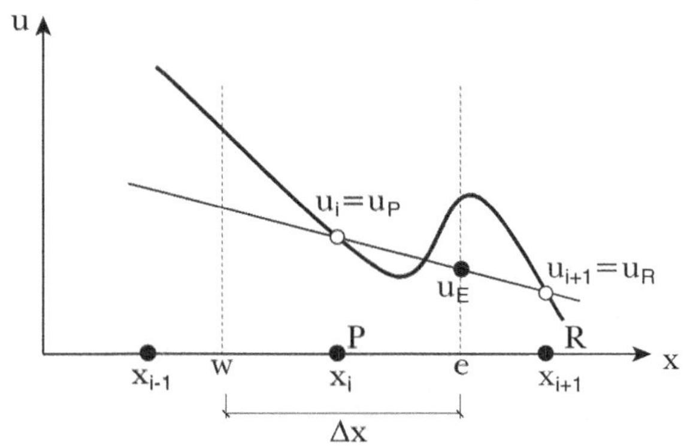

Figure 5.6 Interpolation at point *P*.

5.2.3.2 Central Differences

A central differences scheme (CDS) is also used for the approximation of the convective fluxes. In case of the CDS approach u is linearly interpolated (see Figure 5.6) and equation (5.24) applies for interpolation. The interpolation factor γ_e is defined by equation (5.23).

5.2.4 Discretization of diffusive fluxes

The approximation of the diffusive fluxes, F^D is done by approximating the normal derivative of u at the CV faces. These values are expressed using the values of u in the CV centers. For example on the east face of the CV, Ae, a simple difference formula to express $(\partial u/\partial x)_e$ is:

$$\left(\frac{\partial u}{\partial x}\right)_e \approx \frac{u_R - u_P}{x_R - x_P} \tag{5.25}$$

Equation (5.25) is a central difference formula.

5.2.5 Evaluation of the time derivative

Time derivatives are approximated using fractional steps (or finite differences):

$$\frac{\partial}{\partial t}\int_V u dV\Bigg|_{Vi} \cong V_i \cdot \frac{u_i^{n+1} - u_i^n}{\Delta t} \tag{5.26}$$

5.2.6 Boundary conditions

Boundary conditions can be specified in a number of ways; Dirichlet type, when the unknown function u is prescribed; Neumann type, when the flux of the unknown variable u is prescribed; and a combination of the previous two. Boundary conditions are equations that are added to the linear system of equations obtained over the computational domain in order to have the same number of equations as the number of unknowns.

Consider the CV, as defined in Figure 5.3b. If u is given as a prescribed value on the w side of the CV, as
$u_w = u_{wBC}$ then the convective flux at the boundary is approximated as:

$$F_l^C \approx m_w u_w = m_w u_{wBC} \tag{5.27}$$

Equation (5.27) is used directly into equation (5.20).

The diffusive flux through the boundary is calculated using the same approach as:

$$\left(\frac{\partial u}{\partial x}\right)_w \approx \frac{u_P - u_w}{x_P - x_w} = \frac{u_P - u_{wBC}}{x_P - x_w} \tag{5.28}$$

The algebraic system of equations obtained when the FVM approach is used has a unique solution only if the boundary conditions at all boundaries of the computational domain are defined.

5.2.7 Solving algebraic system of equations

By summing up all elements of equation (5.19) a system of equations is obtained, which gives the values of the unknown function at all center nodes of the CV-s. Replacing the fluxes, temporal integrals and boundary conditions in equation (5.19) for one of the control volumes V_i ($i = 1, \ldots, N$) of the computational domain, the following equation is obtained:

$$a_P^i u_P^i - \Sigma_c a_c^i u_c^i = b_P^i \quad for\, all\ i = 1,\ldots,N \tag{5.29}$$

where c represents the number of all cells around the CV, which has as a center point P. If equation (5.59) is written for each of the control volumes V_i ($i = 1, \ldots, N$), there are N equations formed because there are a total of N discretization nodes.

Schafer *et al.* (1993) shows how such a system of type (5.29) is obtained and solved for a one- and two-dimensional advection diffusion equation, in steady state (no time term).

Consider the steady one-dimensional advection-diffusion equation defined on domain [0, L], discretized into N control volumes (see Figure 5.7). Using a second-order central differencing scheme, the discrete equations of the phenomena, after computing all fluxes are:

$$a_P^i u_P^i - a_R^i u_R^i - a_w^i u_w^i = b_P^i \tag{5.30}$$

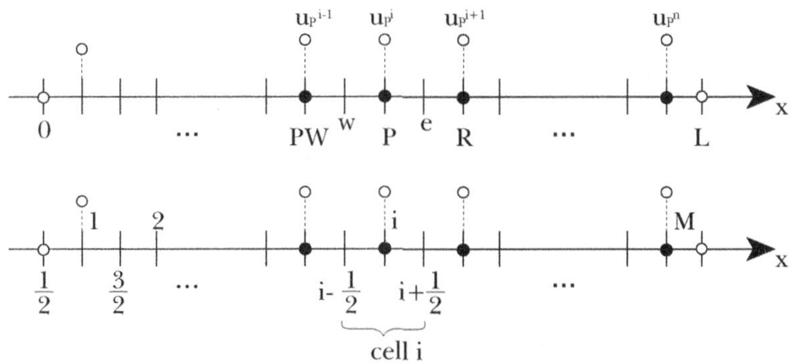

Figure 5.7 Discretization of the 1D computational domain.

Values of the function over each CV, as defined in Figure 5.7, are:

$$u_w^i = u_P^{i-1} \quad \textit{for all} \;\; i-2,\ldots,N \tag{5.31a}$$

$$u_R^i = u_P^{i+1} \quad \textit{for all} \;\; i-1,\ldots,N-1 \tag{5.31b}$$

which leads to a linear system of equations, that can be represented in matrix form as:

$$\underbrace{\begin{bmatrix} a_P^1 & -a_R^1 & & \cdots & & \cdot \\ -a_w^2 & a_P^2 & -a_R^2 & & & 0 \\ & & \cdot & \cdot & \cdot & \\ \vdots & & -a_w^i & a_P^i & -a_R^i & \vdots \\ & & \cdot & \cdot & \cdot & \\ 0 & & \cdots & & & -a_R^{N-1} \\ & & & & -a_w^N & a_P^N \end{bmatrix}}_{A} \underbrace{\begin{bmatrix} u_P^1 \\ \cdot \\ u_P^{i-1} \\ u_P^i \\ u_P^{i+1} \\ \cdot \\ u_P^N \end{bmatrix}}_{u} = \underbrace{\begin{bmatrix} b_P^1 \\ b_P^2 \\ \cdot \\ \cdot \\ b_P^i \\ \cdot \\ \cdot \\ b_P^N \end{bmatrix}}_{u} \tag{5.32}$$

which can be solved to determine the values of u at each nodal point.

If the two-dimensional case of the steady advection-diffusion equation is considered over a computational domain as the one represented in Figure 5.8, then equations (5.19) are of the form:

$$a_P^{i,j} u_P^{i,j} - a_E^{i,j} u_E^{i,j} - a_w^{i,j} u_w^{i,j} - a_A^{i,j} u_A^{i,j} - a_N^{i,j} u_N^{i,j} = b_P^{i,j} \tag{5.33}$$

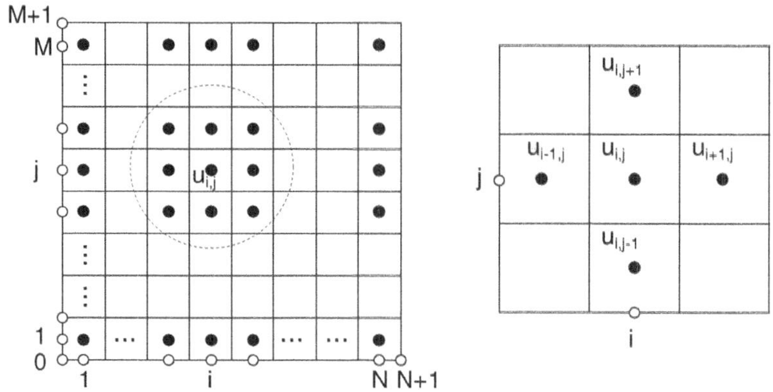

Figure 5.8 Discretization of the 2D computational domain.

where $i = 1, \ldots, N$ are the discretization intervals on the x axis; and $j = 1, \ldots, M$, are the discretization intervals on the y axis. Using the notations from Figure 5.8, the matrix A of the system is:

$$A = \begin{bmatrix} a_P^{1,1} & -a_N^{1,1} & \cdot & 0 & \cdot & -a_R^{1,1} & & & \\ -a_A^{1,2} & \cdot & \cdot & & & & & \cdot & 0 \\ \cdot & \cdot & \cdot & \cdot & & & & & \\ 0 & \cdot & \cdot & & \cdot & \cdot & & & -a_R^{N-1,M} \\ \cdot & & \cdot & & \cdot & \cdot & \cdot & & \cdot \\ -a_w^{2,1} & & & & \cdot & \cdot & & & 0 \\ & & & & & \cdot & \cdot & \cdot & \cdot \\ & 0 & & & & & \cdot & \cdot & -a_N^{N,M-1} \\ & & & -a_w^{N,M} & \cdot & 0 & \cdot & -a_A^{N,M} & a_P^{N,M} \end{bmatrix} \tag{5.34}$$

5.3 EXAMPLE OF ADVECTION-DIFFUSION EQUATION IN 1D

The application of the FVM approach is illustrated here on the one-dimensional advection diffusion equation. The domain of computation for the problem is L and the discretization of the domain is illustrated in Figure 5.9. In order to make the demonstration easier the numbering of CV sides along with the CV center nodes is done as a multiple of number 2. The CV sides are shown as ratios, while the CV center nodes are the numbers $i = 1, \ldots, M$ (see Figure 5.9). The domain has M number of equally spaced volume cells. The width of a cell is Δx, which corresponds with the volume of the CV, because it is a one dimensional problem.

The advection diffusion equation to be solved over the one-dimensional domain is:

$$\frac{\partial u}{\partial t} + a\frac{\partial u}{\partial x} = \alpha\frac{\partial^2 u}{\partial x^2} \tag{5.35}$$

Equation (5.35) has two types of fluxes; the advective flux, which is noted as $F = au$, and the diffusive fluxes noted as $D = \alpha u$. These fluxes are written at any point $x_{j+½}$.

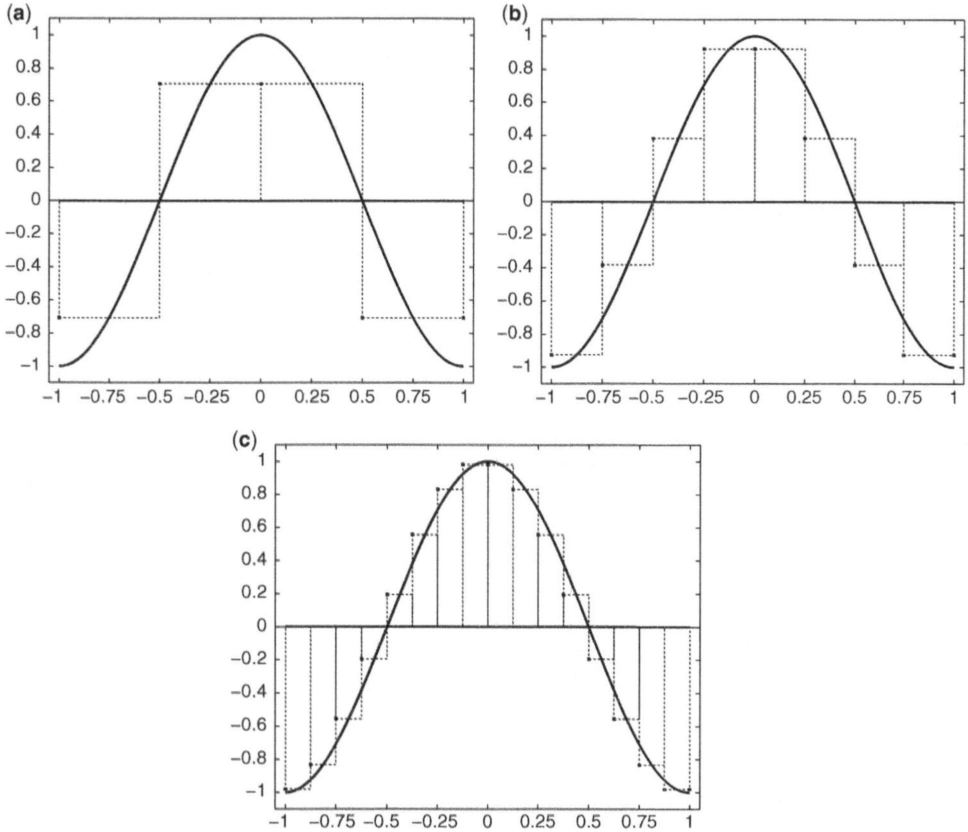

Figure 5.9 Values of the unknown function over an equally spaced grid using a constant function approximation and a discretization of: (a) 4 cells, (b) 8 cells, (c) 16 cells.

Equation (5.35) integrated reduces to:

$$\frac{du_i}{dt} = -\frac{F_{i+\frac{1}{2}} - F_{i-\frac{1}{2}}}{\Delta x} + \frac{D_{i+\frac{1}{2}} - D_{i-\frac{1}{2}}}{\Delta x} \tag{5.36}$$

The evaluation of the terms F and D in equation (5.36) is done at the cell edges of the control volumes.

Before time integration takes place the fluxes are evaluated using the step by step procedure. Values of function u are written as a polynomial representation, as detailed by equation (5.9):

$$u = \sum_{n=0}^{p} a_n \xi^n, \quad \text{where } \xi = \left(\frac{x - x_i}{\Delta x_i}\right) \tag{5.37}$$

In equation (5.37) a_n are $P + 1$ coefficients that are determined from $P + 1$ conditions on the averages of the polynomials over multiple cells, as shown by equation (5.10). In equation (5.37) the coordinate system is centered to the cell and scaled so that the cell length is from –1/2 to 1/2.

Applying equation (5.11) coefficients are computed as:

$$A_{m,n} = \int_{x_{m-\frac{1}{2}}}^{x_{m+\frac{1}{2}}} \left(\frac{x - x_{i-\frac{1}{2}}}{\Delta x_i} \right)^{n-1} dx, \quad for\ n = 1,2,\ldots,P+1 \tag{5.38}$$

$$= \frac{\Delta x_i}{n} \left[\left(\frac{x_{m+\frac{1}{2}} - x_{i-\frac{1}{2}}}{\Delta x_i} \right)^n - \left(\frac{x_{m-\frac{1}{2}} - x_{i-\frac{1}{2}}}{\Delta x_i} \right)^n \right] \tag{5.39}$$

$$= \Delta x_i \frac{\xi^n_{m+\frac{1}{2}} - \xi^n_{m-\frac{1}{2}}}{n}, \quad with\ \xi_{m\pm\frac{1}{2}} = \frac{x_{m\pm i}}{\Delta x_i} \tag{5.40}$$

Values of $A_{m,n}$ are determined and the system of equation (5.38) is solved. Different polynomial representations for the function u are considered and shown bellow, for an initial condition of a cosine function.

5.3.1 Constant unknown function

The constant value of u, as shown in Figure 5.4, is the simplest case, in which the approximation is given by the relation:

$$u = a_0 \tag{5.41}$$

In such a case a_0 is the only unknown coefficient, hence only one constraint has to be fulfilled:

$$\int_{x_{i-\frac{1}{2}}}^{x_{i+\frac{1}{2}}} a_0 dx = u_i \Delta x_i \tag{5.42}$$

Equation (5.40) states that the volume integral over the control volume is equal with the function value times the control volume length, hence:

$$a_0 = u_i \tag{5.43}$$

Two approximations are available for (5.41), either the left value of u, in cell i, or the right value of u in cell $i + 1$. The flux is advective, hence the information usually is traveling from upstream to downstream, therefore, the solution is:

$$u_{i+\frac{1}{2}} = \begin{cases} u_i & if\ a_{i+\frac{1}{2}} \geq 0 \\ u_{i+1} & if\ a_{i+\frac{1}{2}} < 0 \end{cases} \tag{5.44}$$

The diffusive term cannot be computed when constant functions are used because the derivative of a constant is zero.

5.3.2 Linear variation approximation of the unknown function

A better approximation than the constant one is the linear variations of the unknown function over two adjacent control volume cells. If a linear approximation is considered for u then:

$$u = a_0 + a_1\xi \tag{5.45}$$

In equation (5.45) there are two unknown coefficients, a_0 and a_1, hence two conditions are set in order to determine them. Consider two consecutive control volumes i and $(i + 1)$ so that $x_{i-\frac{1}{2}} \le x \le x_{i+\frac{3}{2}}$; hence $-\frac{1}{2} \le \xi \le \frac{3}{2}$. The two necessary conditions to find a_0 and a_1 s are:

$$\int_{-\frac{1}{2}}^{\frac{1}{2}} (a_0 + a_1\xi)d\xi = u_i \tag{5.46}$$

$$\int_{\frac{1}{2}}^{\frac{3}{2}} (a_0 + a_1\xi)d\xi = \frac{\Delta x_{i+1}}{\Delta x_i} u_{i+1} \tag{5.47}$$

After performing the integration, the following system of equations is obtained:

$$\begin{pmatrix} 1 & 0 \\ 1 & 1 \end{pmatrix}\begin{pmatrix} a_0 \\ a_1 \end{pmatrix} = \begin{pmatrix} u_i \\ \dfrac{\Delta x_{i+1}}{\Delta x_i} u_{i+1} \end{pmatrix} \tag{5.48}$$

For an equally spaced discretization, it yields:

$$a_0 = u_i, \quad a_1 = u_{i+1} - u_{i-1} \tag{5.49}$$

The linear interpolation of u over the two adjacent cells gives:

$$u = u_i + (u_{i+1} - u_i) \tag{5.50}$$

hence

$$u = (1 - \xi)u_i + \xi u_{i+1}, \quad for\ -\frac{1}{2} \le \xi \le \frac{3}{2} \tag{5.51}$$

The unknown function u at the edge of the control volume is obtained for values of $\xi = \pm\frac{1}{2}$.

5.3.3 Parabolic variation approximation of the unknown function

Considering an equally spaced discretization grid and a parabolic interpolation of the unknown function u, yields

$$u = \left[\frac{-u_{i+1} + 26u_i - u_{i-1}}{24}\right] + \left[\frac{u_{i+1} - u_{i-1}}{2}\right]\xi + \left[\frac{u_{i+1} - 2u_i + u_{i-1}}{2}\right]\xi^2 \tag{5.52}$$

The interpolation is centered on the control volume cells $(i - 1)$, i and $(i + 1)$, and as in previous cases:

$$-\frac{3}{2} \le \xi \le \frac{3}{2}, \quad \xi = \frac{x - xi}{\Delta x} \tag{5.53}$$

At the edge of the control volume, where $x_{i+1/2}$ and $\xi = 1/2$ the value of the unknown function u is:

$$u_{i+\frac{1}{2}} = \frac{-u_{i-1} + 5u_i + 2u_{i+1}}{6} \tag{5.54}$$

In order to assess the how good are the different forms of representation of the unknown function an example is considered, for which the analytical solution is known. The difference between the two results, the analytical and numerical solution, is the error, which depends on the discretization. The more refined discretization, the smaller the error.

5.3.4 Error of the approximation

Let u be $\cos \pi x$ over the interval $-1 \leq x \leq 1$, as the initial value. The advection equation solution at all time steps is the same as the initial value as it was demonstrated in Chapter 2. The advection equation transfers the initial profile unchanged along x axis.

The result of the approximation using constant, linear and parabolic representations of the function u are shown in Figures 5.9, 5.10 and 5.11, respectively, and the error at each cell edge is given by the height difference between the solid curve (analytical solution) and the square symbol.

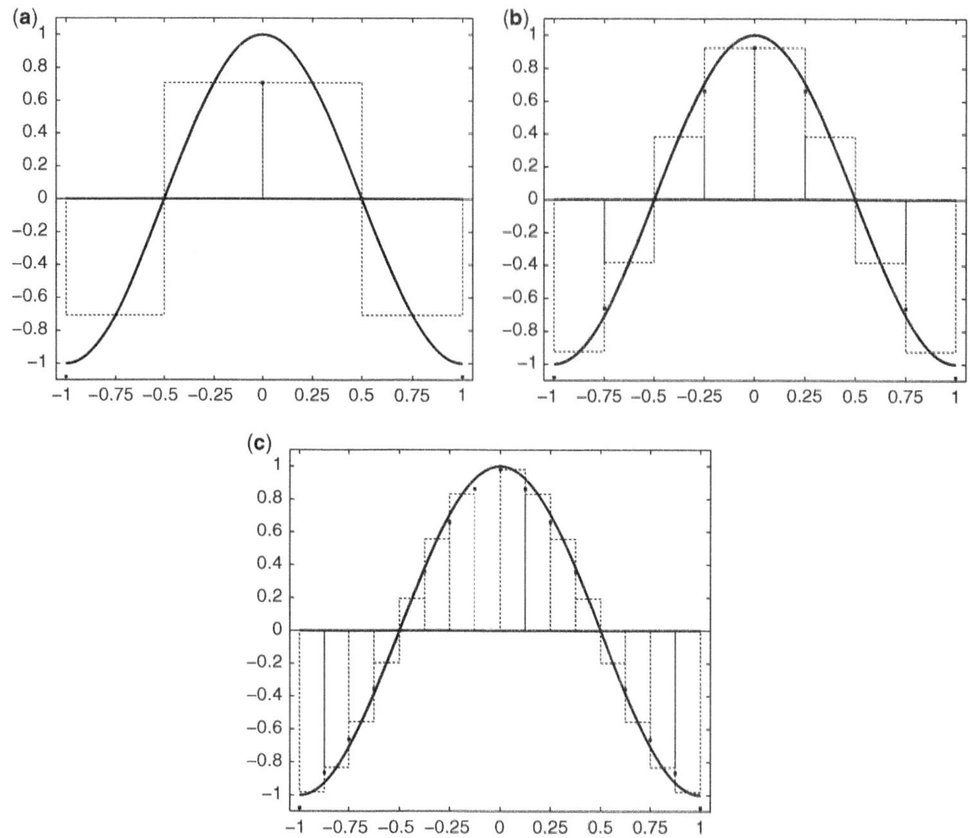

Figure 5.10 Values of the unknown function over an equally spaced grid using a linear function approximation and a discretization of: (a) 4 cells, (b) 8 cells, (c) 16 cells.

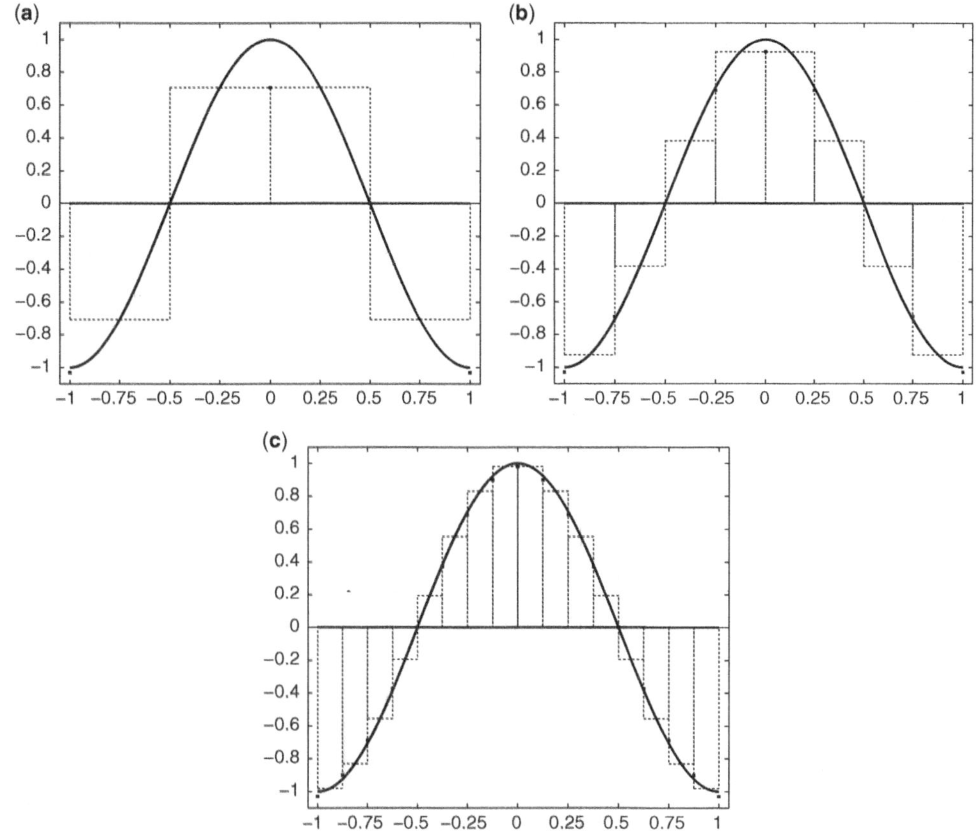

Figure 5.11 Values of the unknown function over an equally spaced grid using a parabolic approximation and a discretization of: (a) 4 cells, (b) 8 cells, (c) 16 cells.

In Figure 5.9 the representation of the solution using constant approximation and a number of 4, 8, and 16 discretization cells is given. The solid curved line is the exact solution, while the dotted lines represent the approximate solution. Similarly, Figures 5.10 and 5.11 shows the representation of the solution for the case of linear and parabolic approximation, respectively. The representation of the linear approximation is done over two cells, while the parabolic one is represented over three cells.

Analysing the results it can be noted that cosine wave is not well represented in case that a coarse discretization grid is used along with a constant approximation for the unknown function. In case that a linear and/or parabolic approximation is used with a finer discretization grid, then the solution is better, however there are still several values that are out of bound.

Chapter 6

Properties of numerical methods

6.1 PROPERTIES OF NUMERICAL METHODS

Each solution method for an ODE or a PDE using numerical approaches gives an approximate value for the unknown function. The errors in the solution come from the way the derivatives are expressed on the discrete parts of the domain. The main question that arises is how good the approximation is? how close is the computed solution to the analytical solution of the differential equation. This can be mathematically achieved by imposing the condition that the numerical solution converges to the exact analytical solution, however this is not an easy task because the analytical solution is in general not possible to be determined. The study of the convergence is actually done by looking at the error behaviour. The study is based on the theory of truncation error (the error between the analytical and approximate solution), which states that refinement of the discretization grid gives better numerical solution (more accurate) (Abbott & Basco, 1989; Fletcher, 1991; Guinot, 2003).

In practice an indirect procedure is used, by imposing that the system of algebraic equations obtained through numerical solution approach, to be consistent with the governing differential equation. In addition the numerical solution should be stable. Consistency and stability ensures convergence. This is stated by Lax theorem of convergence, which is presented bellow.

The process of discretization and ensuring convergence of numerical solution to the analytical solution, as defined schematically by Fletcher (1991), is presented in Figure 6.1.

6.1.1 Convergence

A solution obtained by a numerical scheme is by definition *convergent* if it is closer and closer to the analytical (exact) solution of the differential equation (ODE or PDE) when the grid size of the discretization tends to zero. The difference between the exact solution and the value of the solution obtained with numerical approximation is called *solution error*.

There are two different meanings of convergence in numerical analysis; if the discretized interval is getting finer and finer after discretizing the continuous problems, the solution is convergent to the true solution; and for an iterative scheme, convergence means the iteration gets closer and closer to the true solution as computation progresses.

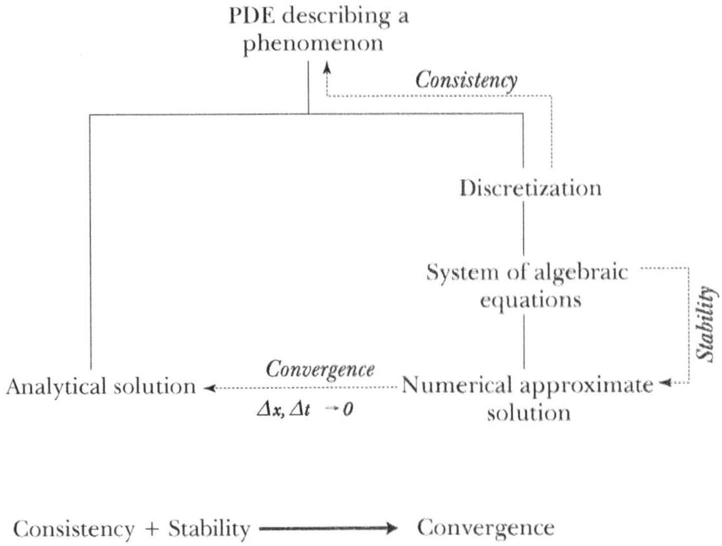

Figure 6.1 Relationship between convergence, stability and consistency.

In order for an approximate solution to converge to the exact solution of a discretised equation (ODE or PDE) as the time and/or space step sizes goes to zero, two conditions must be met: consistency and stability.

6.1.2 Consistency

A solution obtained with a numerical scheme is *consistent* if it gives a correct approximation of the differential equation as the computational, of time and/or space, step is decreased, which means that local truncation error goes to zero as step size go to zero. Consistency is usually verified using Taylor Series expansion. The analysis of the consistency of a scheme is done for each particular studied scheme, and it is further presented in this chapter for the schemes that were introduced in Chapters 4 and 5.

6.1.3 Stability

Stability refers to the magnitude of the change of the error. A scheme is stable if any initially finite perturbation remains bounded as time grows. A scheme is stable if initial errors or small errors at any time remain small while computation progresses. It is unstable if initial errors or small errors at any time get larger and larger, or eventually get unbounded, as computation advances in time.

Though it may seem a very simple issue, it is interesting to mention that stability is an important property of the numerical methods which is not easy to notice and has not even been discovered by the one who defined the finite difference method for the first time, L.F. Richardsond in his book on weather predictions using numerical methods, which was issued in 1922. The issue of stability was noticed by Courant, Friedrichs and Levy in late twenties (hence the name CFL condition); was well defined by von Neumann in 1940's and rigorously defined by Lax and Rychtmyer in 1950's.

Numerical methods can be classified in absolutely stable, conditionally stable and unstable.

Stability verification is more complex than consistency. Several methods are available to carry out stability analysis:

- Matrix method, which is based on finding the eigen values of the matrix representation of the solution given by the numerical scheme. It is used mainly for FDM method.
- Fourier method, or also known as von Neumann analysis, in which the solution given by the numerical method is expressed as a complex Fourier exponential. The substitution of the Fourier representation of the solution into the original equation allows for the analysis of growth or decay of the error. The method however is limited to linear equations.
- Domain of dependence, in which the domains of dependence of the analysed differential equation and the numerical representation of the solution are compared.

6.1.4 Lax's theorem of equivalence

For the restricted class of linear problems Lax's theorem states the relationship between consistency, stability and convergence for finite difference approaches, however the theorem is applicable for any numerical approach that involves discretization of the computational domain and representation of the differential equation on such a domain (Richtmyer & Morton, 1967). The theorem states that consistency and stability are a necessary and sufficient condition to obtain convergence. Any of the two conditions alone are not sufficient to have convergence. Fletcher (1991) states that for the case of fluid problems the theorem cannot always be applied rigorously and should be looked at with care, because fluid problems are non-linear in nature and have mixed boundary and initial conditions.

Lax theorem is a consequence of the analysis of the error between the numerical solution and the exact solution of an equation at a node of the Discretization grid. The analysis of the error, which is expressed as the difference between the exact and the approximate solution, shows that the error at time level n depends on the error introduced at initial time level; on the truncation error; and on the discretization grid. The approximate solution can be stable if the error is bounded and the truncation error is made very small, which is equivalent with the consistency condition.

In practice the above analysis on the bounding of the error is expressed by putting the condition that the growth of the error of the numerical solution should be bounded at all times, which reduces to the so-called Von Neumann stability criterion. The criterion is expressed as:

$$|A| \leq 1 \tag{6.1}$$

where A is called amplification factor and it has the expression:

$$A = \frac{u^{n+1}}{u^n} \tag{6.2}$$

In expression (6.2) u^n is the numerical approximate solution obtained for the unknown function, at time level n of computation, and u^{n+1} at time level $(n + 1)$.

6.2 CONVERGENCE OF FDM SCHEMES

6.2.1 Convergence for ODEs

6.2.1.1 Consistency

In order to analyse consistency of a numerical approximation the truncation error is analysed. In case of ODEs, as defined by equation (2.21), an analysis of the error obtained from time level n of computation to time level $(n + 1)$, gives:

$$TE\ (\Delta t) = u(t^n + \Delta t) - u^{n+1} \tag{6.3}$$

in which u^{n+1} is the approximate solution computed at time level $(n + 1)$ and $u(t^n + \Delta t)$ is the exact value of the function $u(t)$ at the node of the grid where the evaluation of the error is carried out. If $u(t)$ is placed in equation (6.3), the error is a function of only one variable, Δt. Using Taylor series expansion to express the value of $TE(\Delta t)$ around zero, yields:

$$TE\,(\Delta t) = TE\,(0) + \Delta t \left.\frac{d(TE)}{d(\Delta t)}\right]_0 + \frac{(\Delta t)^2}{2!}\cdot\left.\frac{d^2(TE)}{d(\Delta t)^2}\right]_0 + \frac{(\Delta t)^3}{3!}\cdot\left.\frac{d^3(TE)}{d(\Delta t)^3}\right]_0 + \cdots \tag{6.4}$$

According to equation (6.4) any numerical method is a good approximation if the value of the error is as small as possible or equal to zero, hence the smaller the Δt the smaller the error. Equation (6.4) shows the magnitude of the error of the approximation.

A numerical approximation method is defined to be of order p, if the first p terms of the equation (6.4) are zero. Hence a method of order p generates an error for which the first term is:

$$TE(\Delta t) = \frac{(\Delta t)^{p+1}}{(p+1)!}\cdot\left.\frac{d^{p+1}(TE)}{d(\Delta t)^{p+1}}\right]_0 + \cdots \tag{6.5}$$

Consider the weighted method for approximating the solution of an ordinary differential equation. According to the weighted method the approximate solution, as shown in Chapter 4, is:

$$u^{n+1} = u^n + \Delta t \cdot (1-\theta)\cdot f(u^n, t^n) + \Delta t \cdot \theta \cdot f(u^{n+1}, t^{n+1}) \tag{6.6}$$

Equation (6.6) is also known as the multi-step method solution. Ascher and Petzold (1998) defined a general form for this solution as:

$$\sum_{m=0}^{k} \alpha_m \cdot u^{n-m} = \Delta t \cdot \sum_{m=0}^{k} \beta_m \cdot f^{n-m} \tag{6.7}$$

where:

- $\alpha_m \beta_m$ – coefficients of the method;
- u^n – values of the function $u(t)$ at grid point n (see Figure 4.7);
- f^n – value of the derivative of u at grid point n;
- k – the number of past steps applied to solve the ODE.

For the particular case of equation (6.6) the values of the coefficients in equation (6.7) are:

$$\alpha_0 = 1;\quad \alpha_1 = -1;\quad \beta_0 = \theta;\quad \beta_1 = 1-\theta;\quad k = 1; \tag{6.8}$$

The error of approximation for (6.7) at time level n is:

$$TE(\Delta t)\big]_n = \sum_{m=0}^{k} \alpha_m \cdot u^{n-m} - \Delta t \cdot \sum_{m=0}^{k} \beta_m \cdot f^{n-m} \tag{6.9}$$

The order of the method can be determined by analyzing the error, which according to (6.4) is:

$$TE(\Delta t) = \sum_{p=0}^{\infty} C_p \cdot (\Delta t)^{p+1} \cdot \frac{d^p u}{dt^p} \tag{6.10}$$

where the coefficients C_p have the expression:

$$\begin{cases} C_o = \sum_{m=0}^{k} \alpha_m \\ C_p = \sum_{p=0}^{\infty} (-1)^p \cdot \left(\frac{1}{p!} \cdot \sum_{m=1}^{k} m^p \cdot \alpha_m + \frac{1}{(p-1)!} \cdot \sum_{m=0}^{k} m^{p-1} \cdot \beta_m \right) \end{cases} \tag{6.11}$$

Calculation of the above coefficients determines the order of the method, that is, if $C_1, C_2, \ldots, C_p = 0$ and $C_{p+1} \neq 0$ then the method of approximation is of order p. In case of equation (6.6) these coefficients are:

$$\begin{cases} C_o = 1 - 1 = 0 \\ C_1 = 1 - \theta - (1 - \theta) = 0 \\ C_2 = -\frac{1}{2} + (1 - \theta) = 0.5 - \theta \\ C_3 = -\frac{1}{6}(-1) - \frac{1}{2}(1 - \theta) = -\frac{1}{3} + \frac{\theta}{2} \end{cases} \tag{6.12}$$

Based on (6.12) both explicit and implicit methods are first order.

6.2.1.2 Stability

Stability means that if an initial small error is introduced in the initial conditions of the ODE, this should remain bounded, hence the computed solution at the next time level should only slightly change with respect to the original solution. Unstable methods produce oscillatory solutions. If the growth of numerical error is not bounded then the solution is unstable.

Stability condition for an ODE is usually checked using Fourier analysis and the condition expressed by equation (6.2), or what is known as Von Neumann method of analysis.

Ascher and Petzold (1998) show that multi-step methods can be defined using two characteristic polynomials:

$$\rho(Z) = \sum_{m=0}^{k} \alpha_m \cdot Z^{k-m} \tag{6.13a}$$

$$\sigma(Z) = \sum_{m=0}^{k} \beta_m \cdot Z^{k-m} \tag{6.13b}$$

Characteristic polynomials (6.13) are expressed in Z, which is a complex number. For equation (6.6) these two characteristic polynomials have the form:

$$\rho(Z) = Z - 1 \tag{6.14a}$$

$$\sigma(Z) = \theta \cdot Z + (1 - \theta) \tag{6.14b}$$

The numerical solution is stable if the ratio of these two polynomials is less than 1 (Szymkiewicz, 2010), which is equivalent with condition (6.2), hence:

$$A = \frac{\rho(Z)}{\sigma(Z)} = \frac{\rho(e^{i \cdot \varphi})}{\sigma(e^{i \cdot \varphi})} \tag{6.15}$$

where i is the imaginary unit, and φ is the argument (angle) of the complex number, which takes values in the interval $(0, 2\pi)$. Knowing the trigonometric expression, Euler, for a complex number:

$$e^{i \cdot \varphi} = \cos\varphi + i \cdot \sin\varphi \tag{6.16}$$

condition (6.15) is:

$$\begin{aligned} A &= \frac{Z-1}{\theta \cdot Z + (1-\theta)} = \frac{e^{i \cdot \varphi} - 1}{\theta \cdot e^{i \cdot \varphi} + (1-\theta)} = \frac{\cos\varphi + i \cdot \sin\varphi - 1}{\theta(\cos\varphi + i \cdot \sin\varphi) + (1-\theta)} \\ &= \frac{(\cos\varphi - 1) + i \cdot \sin\varphi}{(\theta \cdot \cos\varphi + 1 - \theta) + i \cdot \theta \sin\varphi} \end{aligned} \tag{6.17}$$

A complex number has a real and an imaginary part and A as a ratio of two complex numbers is:

$$A = Re(\theta, \varphi) + i \cdot Im(\theta, \varphi) \tag{6.18}$$

with

$$\begin{cases} Re(\theta,\varphi) = \dfrac{(\cos\varphi - 1)(\theta \cdot \cos\varphi + (1+\theta)) + \theta \cdot \sin^2(\varphi)}{(\theta \cdot \cos\varphi + (1-\theta))^2 + \theta^2 \cdot \sin^2(\varphi)} \\ Im(\theta,\varphi) = \dfrac{(\theta \cdot \cos\varphi + (1-\theta)) \cdot \sin\varphi - \theta \cdot (\cos\varphi - 1)\sin\varphi}{(\theta \cdot \cos\varphi + (1-\theta))^2 + \theta^2 \cdot \sin^2(\varphi)} \end{cases} \tag{6.19}$$

Based on relations (6.19) the region of absolute stability can be represented graphically in the imaginary plane (see Figure 6.2) for different values of θ. Weighting factor θ gives the extent of the region of absolute stability, which is the region for which equation (6.1) holds.

In case of the explicit method $\theta = 0$ and region of stability is very small, while the maximum extent for stability is obtained for $\theta = 1$, which is the case of the implicit method. There is one family of numerical methods for ODEs that are called A-stable methods. By definition (Szymkiewicz, 1993, 1995, 2010) a difference method is considered A-stable if its region of absolute stability covers the entire left pane of the complex plane.

Another important term related to stability analysis of an ODE is the phase error. The phase error is the difference in phase between the analytical and numerical solution and is determined by analysing the argument of the amplification factor. The amplification factor defined by equation (6.2) is expressed in complex form as:

$$A = \frac{u^{n+1}}{u^n} = |A| \cdot e^{i\mu} \tag{6.20}$$

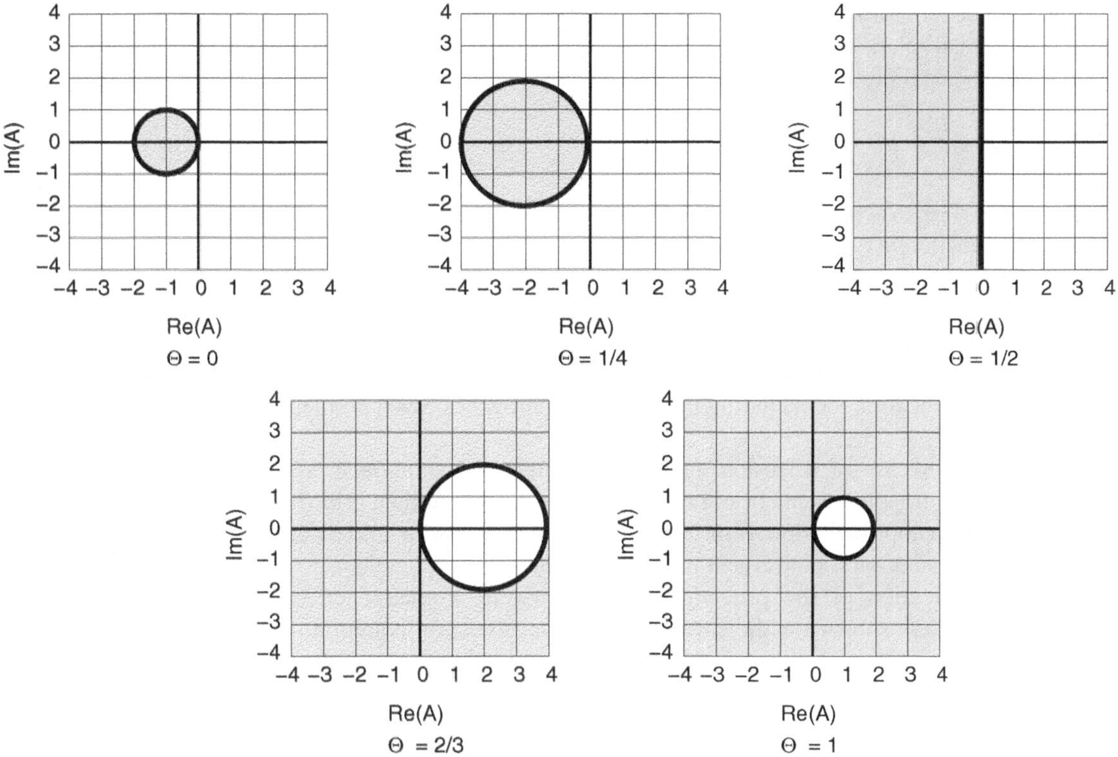

Figure 6.2 Region of stability for multistep methods of solution for ODEs.

In equation (6.18) the term μ is the argument of the amplification factor. If the analytical change of phase for an ODE is noted as μ_a, the ratio of the numerical and analytical phase is called the relative phase error and is expressed as:

$$R = \frac{\mu}{\mu_a} \tag{6.21}$$

Value $R > 1$ means that numerical phase exceeds the analytical phase, hence scheme is called accelerating; while $R < 1$ is called decelerating scheme. Further detailed analysis for accuracy and stability of ODEs can be found in Dahlquist & Bjorck (2007) or LeVeque (2007).

6.2.2 Convergence for PDEs

As it was the case for the ODEs, a numerical solution of a PDE is a good solution if it is convergent to the analytical solution. The approximate solution tends to the exact one, if the grid dimensions are reduced. In case of a one dimensional problem, in the (x, t) computational space this is equivalent with reducing systematically the dimensions of Δt and Δx (i.e., $u_i^n \rightarrow u(x_i, t^n)$, as $\Delta x, \Delta t \rightarrow 0$).

Convergence for PDEs is studied, similarly as for ODEs, by looking at the difference between the exact and the numerical solution of the partial differential equation, at the same discretization grid point. In case of a one dimensional problem the error is expressed as:

$$TE_i^n = u(x_i, t^n) - u_i^n \tag{6.22}$$

in which $u(x_i, t^n)$ is exact value of the function $u(x, t)$ at the grid point (x_i, t^n), and u_i^n is the approximate value of the function $u(x, t)$ at the same grid point.

In general it is very difficult to proof that the numerical solution converges to the exact solution of a PDE, even for simple cases (Fletcher, 1991). The Lax theorem of equivalence is also used for PDEs, though it is applicable just for a restricted class of problems, the ones that are linear. All fluid problems are in general of non-linear nature, hence Lax theorem is used as a necessary condition for convergence and as a useful indication for excluding inconsistent approaches. Each problem has to be treated on a case by case basis.

For example in case of diffusion equation an analytical solution is available, hence a numerical approximation of the solution can be obtained on a successively refined grid. The error of approximation is computed for each approximation. If the error reduces to zero, as the grid discretization is reduced implies that approximate solution is convergent to the analytical solution.

The exemplification of the procedure to demonstrate convergence of a numerical solution to the analytical one is shown for the case of the linear advection equation, as represented by equation (4.43) in Chapter 4. A one dimensional linear advection equation with positive constant advection velocity a is:

$$\frac{\partial u}{\partial t} + a\frac{\partial u}{\partial x} = 0 \tag{6.23}$$

In case of an equally spaced grid, on both x and t directions, the forward time backward space (FTBS) up-wind numerical solution to the equation is given by equation (4.48), as:

$$u_i^{n+1} = Cr \cdot u_{i-1}^n + (1 - Cr) \cdot u_i^n \tag{6.24}$$

Recall that Cr is the Courant number:

$$Cr = a \cdot \frac{\Delta t}{\Delta x} \tag{6.25}$$

Consider that $u(x, t)$ represents the water depth in a 10 km long canal, which has an initial water level of $u_0 = 5\ m$. Based on the canal geometry the advection velocity is $a = 0.5$ m/s.

For the given initial condition of the form:

$$u(x=0,t) = \begin{cases} u_0 \cdot \dfrac{t}{t^m}, & 0 \le t < t^m \\ u_0 \cdot \left(2 - \dfrac{t}{t^m}\right), & t^m \le t < 2t^m \\ 0, & t \ge 2t^m \end{cases} \tag{6.26}$$

equation (6.23) has an exact solution as shown in Figure 6.3. The exact solution at a later moment in time and space is the same as the initial condition, that is, initial condition is, advected unchanged towards downstream. For different values of grid discretization, such that Courrant number is kept constant, $Cr = 0.5$, the result is presented in Figure 6.3. Comparing numerical solutions with the analytical solution, it is noticed that the numerical solution tends to the analytical solution as the grid dimensions are reduced, which shows that the up-wind FTBS scheme converges to the analytical solution.

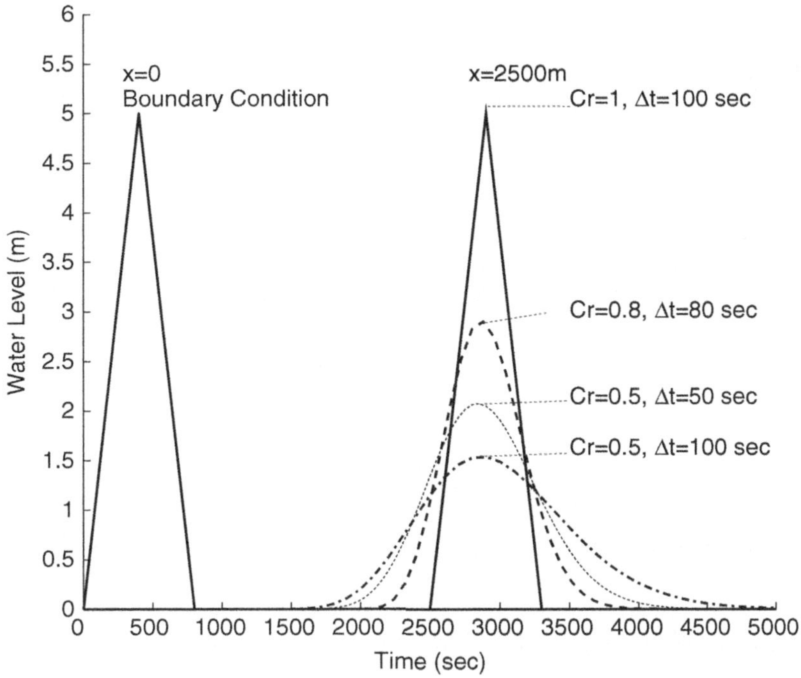

Figure 6.3 Linear advection equation's initial condition, exact solution and numerical approximations by the FTBS up-wind scheme.

The above demonstration shows that for the particular case of advection equation a particular numerical scheme converges to the analytical solution, however this is not easy to do for other types of equations.

6.2.2.1 Consistency for PDE

A numerical approximation of a PDE is consistent with the equation itself if in the limit, when the discretization's dimensions tend to zero, the approximation becomes the PDE at every node of the grid. Consistency is tested by replacing the values of the unknown function by their Taylor series expansion around the considered node. The equation obtained by using Taylor series expansions is not the same as the original PDE, it contains the original equation plus additional terms (reminder). The condition of consistency requires that the structure of the reminder is as such that it tends to zero as the grid is refined.

In case of the numerical scheme (6.24) for the linear advection equation Taylor series expansion of the point values u_{i-1}^n, and u_i^n around point (i, n) are:

$$u_{i-1}^n = u_i^n - \Delta x \left.\frac{\partial u}{\partial x}\right|_i^n + \left.\frac{\Delta x^2}{2}\frac{\partial^2 u}{\partial x^2}\right|_i^n - \left.\frac{\Delta x^3}{6}\frac{\partial^3 u}{\partial x^3}\right|_i^n + \cdots \tag{6.27}$$

$$u_i^n = u_i^n + \Delta t \left.\frac{\partial u}{\partial t}\right|_i^{n+1} + \left.\frac{\Delta t^2}{2}\frac{\partial^2 u}{\partial t^2}\right|_i^{n+1} - \left.\frac{\Delta t^3}{6}\frac{\partial^3 u}{\partial t^3}\right|_i^{n+1} + \cdots \tag{6.28}$$

Substitution in equation (6.24) gives:

$$\frac{\partial u}{\partial t} + a\frac{\partial u}{\partial x} = TE(\Delta x, \Delta t) \tag{6.29}$$

where $TE(\Delta x, \Delta t)$ is the reminder term (or the truncation error term):

$$TE(\Delta x, \Delta t) = -\frac{\Delta t}{2}\cdot\frac{\partial^2 u}{\partial t^2} - \frac{a\Delta x}{2}\cdot\frac{\partial^2 u}{\partial x^2} + \frac{\Delta t^2}{6}\cdot\frac{\partial^3 u}{\partial t^3} - \frac{a\Delta x^2}{6}\cdot\frac{\partial^3 u}{\partial x^3} + \cdots \tag{6.30a}$$

Rearranging (6.30*a*) gives:

$$\frac{\partial u}{\partial t} + a\frac{\partial u}{\partial x} = \underbrace{\underbrace{(1 - Cr)\frac{a\Delta x}{2}\cdot\frac{\partial^2 u}{\partial x^2}}_{TE1} + \underbrace{\mathrm{O}(\Delta x^2)}_{TE2} + \underbrace{\mathrm{O}(\Delta t^2)}_{TE3}}_{TE(\Delta x, \Delta t)} \tag{6.30b}$$

where $\mathrm{O}(\Delta x^2)$ and $\mathrm{O}(\Delta t^2)$ are terms that contain higher order derivatives starting with third order derivatives with respect to x and t.

In equations (6.29) and (6.30) the indices i and n are omitted, however expressions are valid for the discretization node (i, n). The term $TE(\Delta x, \Delta t)$ is dependent on the selected approximation method. As the dimensions of the grid discretization tend to zero $TE(\Delta x, \Delta t)$ tends to zero and consequently equation (6.29) becomes the original PDE (equation (6.23)).

The approach demonstrated to prove consistency for the linear advection equation can be applied for any numerical scheme.

6.2.2.2 Stability for PDE

A numerical solution for a PDE should not only be consistent with the discretized PDE, it should also be stable. Stability is the property of a numerical method that keeps any initial error bounded. Errors may be due to boundary conditions, parameters, or due to round-off errors of computers. A numerical method is considered stable if a small variation in the input data causes small variation in the final result. If a method is not capable to keep initial errors bounded and they are growing in time and space then the method is called unstable.

Considering again the example of the advection equation and further approximating the solution, using values for Δt and Δx such that $Cr > 1$ (see Figure 6.4), it is noticed that solution is affected by oscillations and does not represent accurately the exact solution. Oscillations increase in amplitude over time, as well as in space.

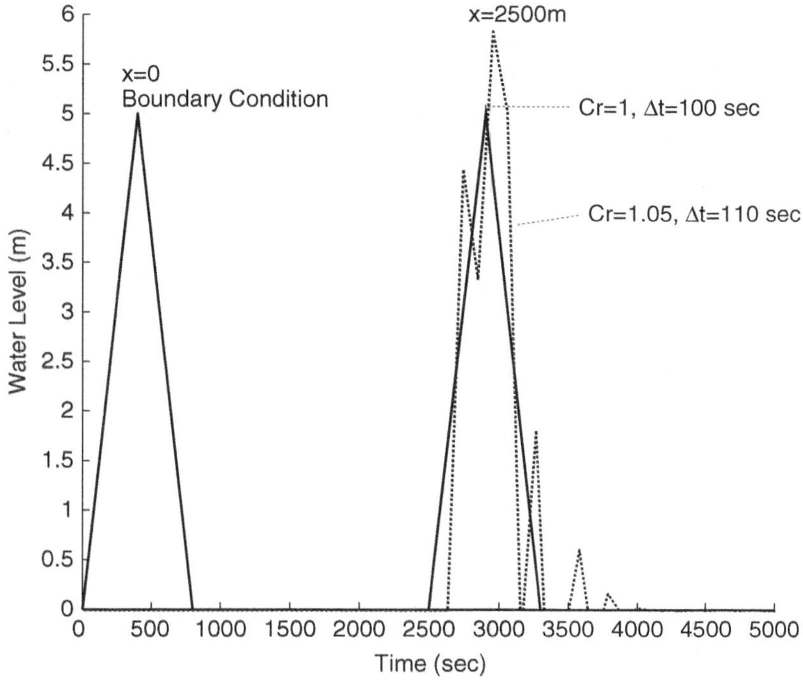

Figure 6.4 Linear advection equation's numerical approximations by the FTBS up-wind scheme for Courant numbers greater than 1.

Conditions of stability of the advection equation are investigated bellow. According to equation (6.22) each approximated value of the unknown function at a discretization node is a difference between the exact value and the error. For the three terms in equation (6.24) it yields:

$$\begin{cases} u_i^{n+1} = u(x_i, t^{n+1}) - TE_i^{n+1} \\ u_i^n = u(x_i, t^n) - TE_i^n \\ u_{i-1}^n = u(x_{i-1}, t^n) - TE_{i-1}^n \end{cases} \tag{6.31}$$

Relations (6.31) are replaced into equation (6.24) to obtain:

$$u(x_i, t^{n+1}) - TE_i^{n+1} = Cr(u(x_{i-1}, t^n) - TE_{i-1}^n) + (1 - Cr)(u(x_i, t^n) - TE_i^n) \tag{6.32}$$

The exact solution of the linear advection equation satisfies relation (6.24) of the numerical scheme:

$$u(x_i, t^{n+1}) = Cr \cdot u(x_{i-1}, t^n) + (1 - Cr) \cdot u(x_i, t^n) \tag{6.33}$$

hence:

$$TE_i^{n+1} = Cr \cdot TE_{i-1}^n + (1 - Cr) \cdot TE_i^n \tag{6.34}$$

Equation (6.34) is written for all the points of the discretization grid. The analysis using von Neumann method implies to expand all errors from time level n as finite complex Fourier series and to analyze how a single component of the series behaves from time level n to time level $(n + 1)$. If each component of the Fourier series is stable then the numerical solution is stable. If one single component of the Fourier series is not stable then the solution is not stable.

Based on the above statements looking at the error term of time level n, TE_j^n expressed in terms of Fourier series (Abbott & Basco, 1989) gives:

$$TE_j^n = \sum_{k=1}^{KF} (Ao \cdot A_k^n) \cdot e^{i \cdot k \cdot m \cdot j \cdot \Delta x} \tag{6.35}$$

(*Note*: Throughout the book indices notation i is used for space discretization. In order to avoid confusion with the unit imaginary number, j is used as an indices notation along x axis, in the Fourier series (6.35).)

In relation (6.35) notations are:

- j: 2,3, ..., M, – index of the discretization nodes,
- i: the imaginary unit,
- n: index of time level,
- k : order of the Fourier component,
- m: wave number,
- Δx: space discretization interval,
- $Ao \cdot A_k^n$: Fourier amplitude of component kth at the time level n, in which Ao is the initial amplitude and A_k^n is the amplification factor from one time level to another,
- KF: the finite number of considered Fourier components.

Equation (6.35) corresponds to a sine wave travelling along x axis with an amplitude (Ao) multiplied by a term (A_k^n) at each time step. Such a wave is called a harmonic component, or Fourier component, or Fourier mode. For the case of linear PDE with constant coefficients, the Fourier modes behave independently one from another, being amplified or damped based on their wave length only.

The relation between wave number (m) and wave length (λ) is:

$$m = \frac{2\pi}{\lambda} \tag{6.36}$$

If parameter N is introduced as being the number of discretization grid intervals available in one wave length:

$$N = \frac{\lambda}{\Delta x} \tag{6.37}$$

and equation (6.36) is multiplied by Δx on both sides then the term φ (the phase of the error) called the *wave number*, is defined as:

$$\varphi = m \cdot \Delta x = \frac{2\pi}{N} \tag{6.38}$$

The shortest waves represented at the considered grid point is $2\Delta x$, while the longest one is ∞, hence $2 \le N \le \infty$, which gives the range of the wave number in the interval $[0, \pi]$.

The error term (6.35), for only one Fourier component, $k = 1$, is:

$$TE_j^n = A^n \cdot e^{i \cdot \varphi \cdot j} \tag{6.39}$$

Substitution of equation (6.39) into (6.34) yields:

$$A^{n+1} \cdot e^{i \cdot \varphi \cdot j} = Cr \cdot A^n \cdot e^{i \cdot \varphi (j-1)} + (1 - Cr) A^n \cdot e^{i \cdot \varphi \cdot j} \tag{6.40}$$

The ratio:

$$A = \frac{A^{n+1}}{A^n} = 1 - Cr + Cr \cdot e^{-i \cdot \varphi} \tag{6.41}$$

is the amplification factor of the k-th component of the Fourier series representation of the error function TE while passing from time level n to time level $(n + 1)$. The error does not increase if the modulus of the amplification factor is less than one, that is, $|A| \leq 1$.

Knowing from (6.16) that $e^{-i \cdot \varphi} = \cos \varphi - i \cdot \sin \varphi$, yields:

$$A = 1 - Cr + Cr \cdot \cos \varphi - i \cdot Cr \cdot \sin \varphi \tag{6.42}$$

Equation (6.42) shows that A is a complex number, hence condition on the modulus is:

$$|A| = \sqrt{(1 - Cr(1 - \cos \varphi))^2 + (Cr \cdot \sin \varphi)^2} \leq 1 \tag{6.43}$$

Considering the trigonometric relationships:

$$\cos \varphi = \cos^2 \frac{\varphi}{2} - \sin^2 \frac{\varphi}{2} \quad \textit{and} \quad \sin \varphi = 2 \sin \frac{\varphi}{2} \cos \frac{\varphi}{2} \tag{6.44}$$

Equation (6.43) gives:

$$|A|^2 = 1 - 4Cr(1 - Cr) \sin^2 \frac{\varphi}{2} \tag{6.45}$$

The wave number φ ranges in the interval $[0, \pi]$. For $\varphi = 0$ the inequality (6.43) is verified $(1 = 1)$. In case $\varphi = \pi$ the maximum value of $\sin^2\left(\frac{\varphi}{2}\right)$ in expression (6.44) is 1. The term $Cr(1 - Cr)$ is a parabola which has a maximum value of 0.25 for $Cr = 0.5$. Hence the amplification factor remains less or equal to 1 as long as:

$$0 \leq Cr \leq 1 \quad \text{or} \quad \Delta \text{t} \leq \frac{\Delta x}{a} \tag{6.46}$$

Equation (6.46) is known as the *Courant-Friederichs-Levy* condition (*CFL condition*) and is valid for hyperbolic equations solved by explicit schemes. The CFL condition is a very important one, however is not a sufficient condition for convergence. It states that a necessary condition for convergence of a numerical scheme to the analytical solution is that the numerical domain of dependence contains the analytical domain of dependence. Courant number is also known to measure the 'numerical speed' of the numerical solution.

The FTBS scheme for the solution of the linear advection equation proofs to be a consistent and a stable scheme, if $Cr < 1$, hence it is a convergent scheme.

6.2.2.3 Numerical diffusion and dispersion

Consistency and stability addressed the issue of convergence, however they are not the only measures for the good fit of a numerical solution with the analytical one. There are other errors that can occur while solving a PDE, which can be of equal importance. For hyperbolic type of equations, such as the case of the advection equation, the solution has the form of a wave, hence wave dynamics are important. It is important that the solutions of such equations are not distorting the wave propagation. The applied numerical solution should not change the wave amplitude nor its phase celerity, which is equivalent with saying that the numerical solution should not generate any dissipation or dispersion of artificial nature. A numerical method, which produces artificial dissipation is called *dissipative* and a method which generates artificial dispersion is called *dispersive.*

Numerical dispersion and dissipation of the solution of hyperbolic and parabolic PDEs are more difficult to overcome as compared to elliptic equations. This is due to the fact that the equations contain advective terms. The effect of numerical diffusion can be noticed on Figure 6.5, which shows the effect of varying numerical parameters on the solution of the advection equation (6.23). In Figure 6.5 a one-dimensional Gaussian shaped initial condition to the advection equation for which $a = 1$ is shown. Numerically the same solution as the analytical one is found for $Cr = 1$, but for $Cr = 0.5$ the initial shape is damped. Numerical diffusion causes amplitude error, and tends to make computed profiles smoother than they should be. In case of transport of the pollutant in a river numerical diffusion leads to an underestimation of peak values in the concentration profile, which is undesirable. Diffusion problems are related to the stability of a scheme and are produced by the first term in the expression of *TE*.

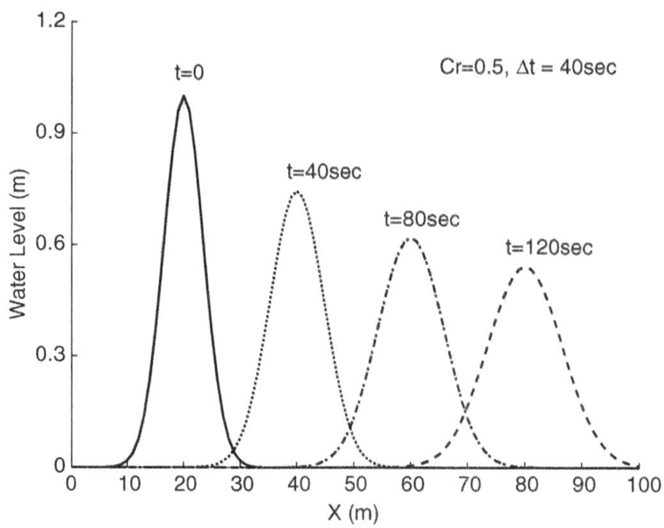

Figure 6.5 Numerical diffusion.

Because numerical diffusion is not a desirable phenomenon, a better scheme to discretise the advection equation is usually used. For example the Preissmann scheme with $\theta = \psi = 1/2$ is second-order accurate, which implies that the first term in the truncation error is zero. The equation is consistent with the following one:

$$\frac{\partial u}{\partial t} + a\frac{\partial u}{\partial x} = k\frac{\partial^3 u}{\partial x^3} \tag{6.47}$$

Equation (6.47) is in fact a dispersion equation. Instead of smoothing profiles artificially, a dispersive scheme tend to make them sharper and create oscillations. This is the numerical dispersion effect. Figure 6.6 shows a typical example of computed profile using a dispersive scheme.

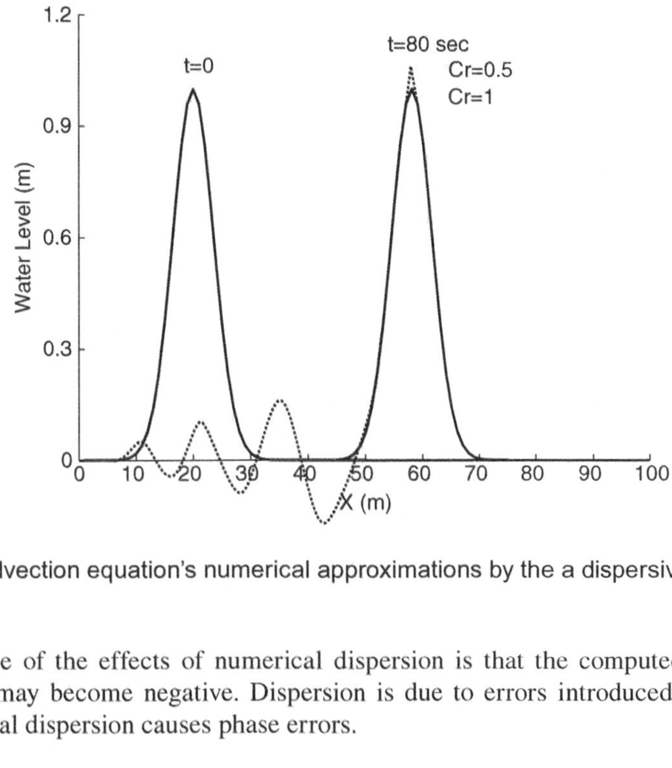

Figure 6.6 Linear advection equation's numerical approximations by the a dispersive scheme.

A typical example of the effects of numerical dispersion is that the computed concentrations of a pollutant in a river may become negative. Dispersion is due to errors introduced when estimating the derivatives. Numerical dispersion causes phase errors.

6.2.3 Amplitude and phase errors

As it can be seen from equation (6.42) the amplification factor (A) for the advection equation is a complex number. Both its amplitude ($|A|$) and its argument (μ) characterise the numerical scheme (Note of caution: φ is the phase of the error related to wave number, not the argument of the amplification factor). The modulus of A shows how the various Fourier modes are being amplified from one time step to the next, while its argument (or phase μ) show how fast the modes are travelling. From equation (6.45) it can be noticed that various Fourier modes are not being amplified in the same way from one time step to the next, because for different wave lengths correspond different values of $m\Delta x$. Similarly Fourier modes do not necessarily travel at the same speed. The graph that shows how the Fourier components are amplified is called amplitude portrait and the one showing at what speed the Fourier components travel is called a phase portrait.

For the FTBS numerical scheme for the advection equation the argument of the amplitude factor given by (6.42) is:

$$\mu = \arg(A) = \arctan\left(\frac{Im(A)}{Re(A)}\right) = \arctan\left(\frac{-Cr \cdot \sin\varphi}{1 - Cr + Cr \cdot \cos\varphi}\right) \tag{6.48a}$$

Following the same approach as for the FTBS scheme the argument of the exact solution results in:

$$\mu_a = \arctan(-Cr \cdot \varphi) = -Cr \cdot m \cdot \Delta x = \frac{2Cr \cdot \pi}{N} \tag{6.48b}$$

Based on (6.48) the ratio of the numerical to analytical phase speed R (see equation (6.21)) is calculated. Amplitude portrait is shown in Figure 6.7 for the advection equation for the FTBS scheme for different values of the Courant number.

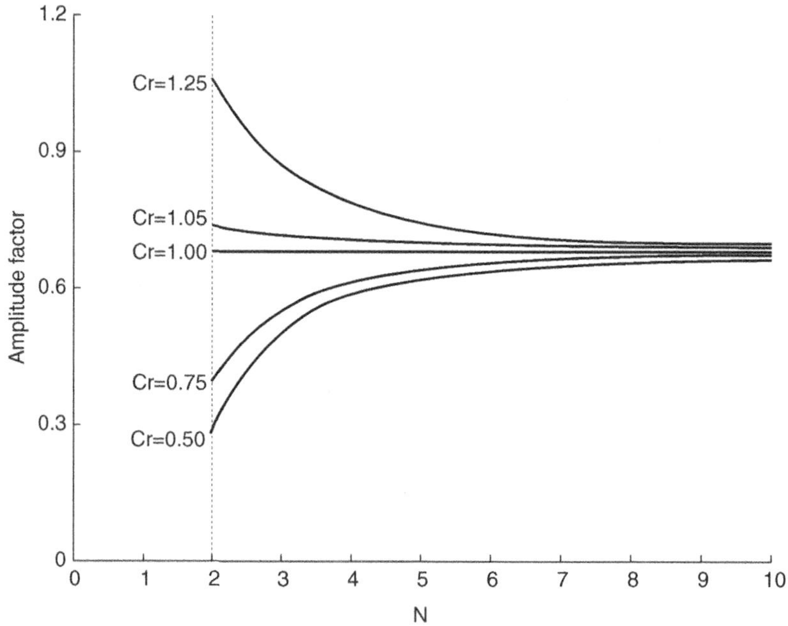

Figure 6.7 Amplitude portrait for advection equation using FTBS numerical approximation.

Figure 6.8 shows the decrease in amplitude of Fourier components for the FTBS scheme after one time step and after 100 time steps, using a value of Courant number of 0.5. Analysing the amplitude decrease, it can be concluded that FTBS scheme has strong numerical diffusion. This is due to the fact the scheme is of first order and truncation errors are large. FTBS scheme is not recommended for the solution of the advection equation, except for cases when boundary conditions are outflows.

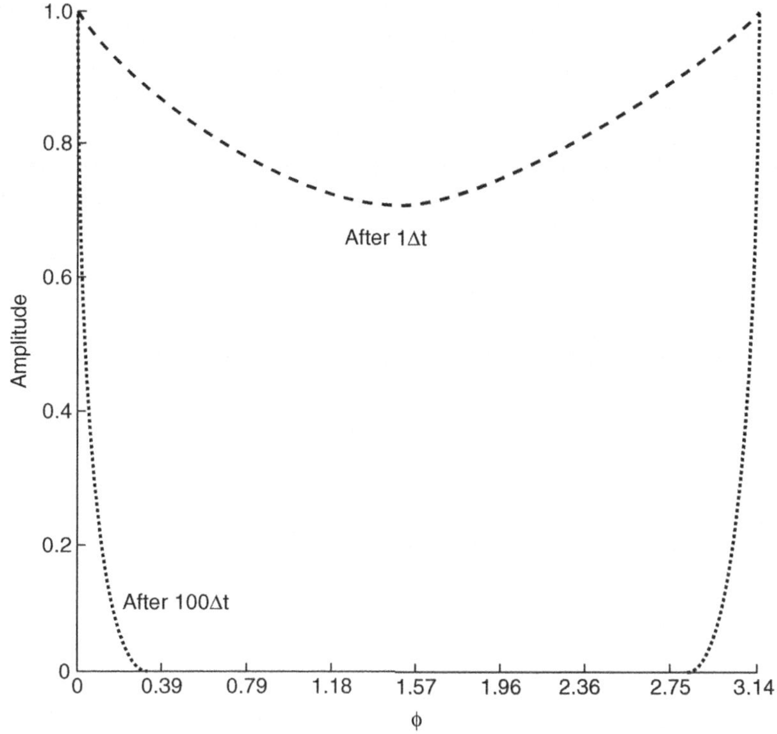

Figure 6.8 Amplification factor of the wave components for advection equation.

6.3 CONVERGENCE OF FVM SCHEMES

6.3.1 Convective fluxes

Convergence of the FVM schemes are analysed for the discretization of convective and diffusive fluxes, as they have been developed in Chapter 5.

The discretization of the convective fluxes in equation (5.19) was done using the upwind and the central schemes.

In case that an upwind scheme is used the Taylor series expansion of the solution of u, at u_E, around the middle point P (see Figure 5.7) gives:

$$u_E = u_P + (x_E + x_P)\left(\frac{\partial u}{\partial x}\right)_P + \frac{(x_E - x_P)^2}{2}\left(\frac{\partial^2 u}{\partial x^2}\right)_P + TE \tag{6.49}$$

In equation (6.49) TE is referred as previously as truncation error. From equation (6.49) it can be seen that the method does not dependent on the computational grid. The interpolation error of the scheme is of 1st order. The first term in the TE of the convective flux F_c^C is:

$$TE_1 = m_e(x_e - x_P)\left(\frac{\partial u}{\partial x}\right)_P \tag{6.50}$$

The error as it is expressed by equation (6.50) is called numerical diffusion (sometimes artificial diffusion), because the term is like a diffusive flux. The advantage of using an upwind scheme is that such a method generates an unconditionally bounded error. The magnitude of the numerical diffusion is given by the coefficient $m_e(x_e - x_p)$. In case that the transport takes place perpendicular to the control volume face, then the approximation of the convective fluxes is good, because the term $(\partial u/\partial x)_P$ is small in such a case. In case that the condition of transport being perpendicular on the face of the control volume is not met then the approximation may be inaccurate and for large values of fluxes (such as large velocities) it is required to use very fine grids in order to obtain good accuracy of the solution. Very fine grids would make the difference $x_e - x_P$ very small.

In case that a central scheme is used for the discretization of convective fluxes then the approximation has an interpolation error of a second order, as it is illustrated by the interpolation in equation (5.12). The second order of the interpolation error is valid for both an equally spaced grid and an unequally spaced grid. Taylor series expansion of u, around the middle point P gives:

$$u(x) = u_P + (x - x_P)\left(\frac{\partial u}{\partial x}\right)_P + \frac{(x - x_P)^2}{2}\left(\frac{\partial^2 u}{\partial x^2}\right)_P + TE \tag{6.51}$$

Evaluating expression (6.51) at x_e and x_E and comparing them leads to:

$$u_e = \gamma_e u_E + (1 - \gamma_e)u_P - \frac{(x_e - x_P)(x_E - x_e)}{2}\left(\frac{\partial^2 u}{\partial x^2}\right)_P + T_H \tag{6.52}$$

which shows that the first term in the error depends quadratically on the grid discretization.

6.3.2 Diffusive fluxes

The analysis of the error magnitude of the diffusive fluxes is done by analysing the difference of the Taylor series expansion at points x_P and x_E, around point x_e. Hence:

$$\left(\frac{\partial u}{\partial x}\right)_e = \frac{u_E - u_P}{x_E - x_P} + \frac{(x_e - x_P)^2 - (x_E - x_e)^2}{2(x_E - x_P)}\left(\frac{\partial^2 u}{\partial x^2}\right)_e - \frac{(x_e - x_P)^3 + (x_E - x_e)^3}{6(x_E - x_P)}\left(\frac{\partial^3 u}{\partial x^3}\right)_e + TE \tag{6.53}$$

Analysing expression (6.53) it can be noticed that for an equally spaced grid the second term of the expression becomes zero, hence a second order error results. In case that the grid is not equally spaced, shows that the term is proportional with the expansion rate of the neighboring grid cell, ξ_e:

$$\frac{(1 - \xi_e)(x_e - x_P)}{2}\left(\frac{\partial^2 u}{\partial x^2}\right)_e, \quad with\ \xi_e = \frac{x_E - x_e}{x_e - x_P} \tag{6.54}$$

which is equivalent with stating that the larger the expansion, the larger the error. This gives modellers the indication on how should unequally spaced grids be build, and that expansion from one grid cell size to another should be done smoothly and with care.

There are other possibilities available for approximating derivatives when using finite volumes, such as forward or backward differencing formulas, however they are hardly used in practice, because of the advantage that central differencing has of being very accurate. Error bounding is not an issue for the diffusive fluxes.

6.4 EXAMPLES

6.4.1 Stability region of a simple ODE

For illustrating the above principles consider the simple ODE and its initial conditions as defined by equation (6.23) below:

$$\begin{cases} \dfrac{du(t)}{dt} = \lambda u(t), \quad for\ t \in [t^0, t^{final}] \\ u(t^0) = u^0 \end{cases} \tag{6.55}$$

In the ODE defined by (6.54), λ may be a complex number (i.e., $\lambda = \lambda_r + i\lambda_i$). Introducing an equally spaced grid with $(N+1)$ nodes of discretization, such that:

$$t^i = t^0 + \Delta t, \quad \Delta t = \frac{t^{final} - t^0}{N} \tag{6.56}$$

and using an explicit discretization approach (Euler's forward finite difference method), the solution of the ODE is given by:

$$u^{n+1} = u^n + \Delta t \cdot \lambda \cdot u^n = u^n(1 + \lambda \cdot \Delta t), \quad n = 0,1,2,\ldots,N \tag{6.57}$$

Equation (6.57) has a recursive character, hence:

$$u^n = u^0(1 + \lambda \cdot \Delta t)^n = u^0 \cdot A^n \tag{6.58}$$

where A is the amplification factor, a complex number of the form:

$$A = 1 + \lambda_r \cdot \Delta t + i \cdot \lambda_i \cdot \Delta t \tag{6.59}$$

As previously shown, in order to obtain the region of stability the condition to be met is:

$$|A| \le 1) \Leftrightarrow |A|^2 = (1 + \lambda_R)^2 + (\lambda_I h)^2 \le 1 \tag{6.60}$$

This condition represents the unit circle centred at $Z = -1$ in the complex plane Z. As long as $\lambda \cdot \Delta t$ is positioned inside the circle the numerical method is stable. This shows that if λ is a negative real number, stability condition requires that

$$\Delta t \le \frac{2}{|\lambda|} \tag{6.61}$$

Such an expression shows that the method is conditionally stable.

6.4.2 Convergence of an ODE: Emptying of a groundwater reservoir

In order to study the accuracy of explicit and implicit schemes for ODEs the example of a cylindrical groundwater reservoir is considered. The volume balance of the reservoir (Figure 6.9) is given by a simple ODE, showing the variation in time of the reservoir water depth.

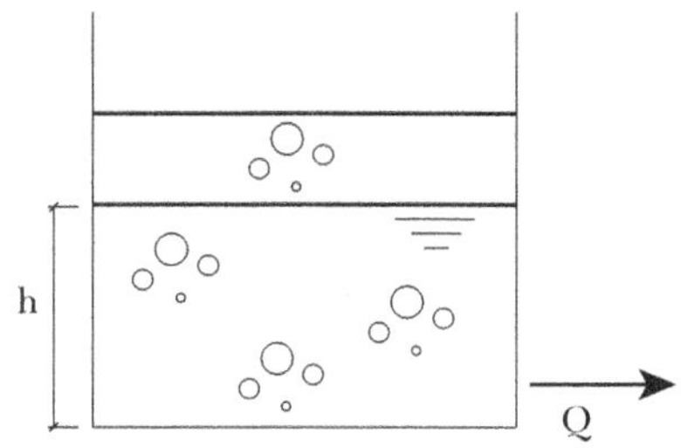

Figure 6.9 Cylindrical ground water reservoir.

Assuming a constant reservoir surface area A, a constant porosity n and an outflow discharge Q defined as linear function of the water level h, gives the following ODE representation for the reservoir:

$$nA\frac{dh}{dt} = -Q \tag{6.62}$$

Eliminating Q, which is a linear function of h, yields:

$$\frac{dh}{dt} = -\alpha h \tag{6.63}$$

where α is the linear reservoir coefficient. Equation (6.62) has the exact solution

$$h = h_0 e^{-\alpha t} \tag{6.64}$$

6.4.2.1 Explicit schemes

An explicit scheme for equation (4.14), written on an equally spaced discretization $(\Delta - t)$ of the independent variable, gives:

$$h^{n+1} = h^n - \alpha \cdot \Delta t \cdot h^n = (1 - \alpha\Delta t) \cdot h^n \tag{6.65}$$

If the numerical solution (6.65) of the equation (6.63) is represented graphically for different (Δt), for a given initial condition of $h = 15$ m (see Figure 6.10), a comparison with the analytical solution (6.64) indicates that the solution decreases too slowly when Δt is large. Moreover this simple example shows that numerical methods are not always accurate, for example for a value of Δt of 22 days, failures in determining the accurate solution may be observed. This can be avoided by choosing the time step in a proper way. Analysis of the accuracy of the method is further done.

Equation (6.65) is an accurate approximation of the equation (6.63) if it behaves in the same way as the function to which it is an approximation, and if it becomes a better approximation as the time-step is reduced. Starting from the first time-step of discretization, when $t = t^o$, equation (6.65) yields:

$$\begin{cases} h^1 = (1 - \alpha\Delta t)h^0 \\ h^2 = (1 - \alpha\Delta t)h^1 = (1 - \alpha\Delta t)(1 - \alpha\Delta t)h^0 = (1 - \alpha\Delta t)^2 h^0 \\ \qquad \dots \\ h^n = (1 - \alpha\Delta t)^n h^0 \end{cases} \tag{6.66}$$

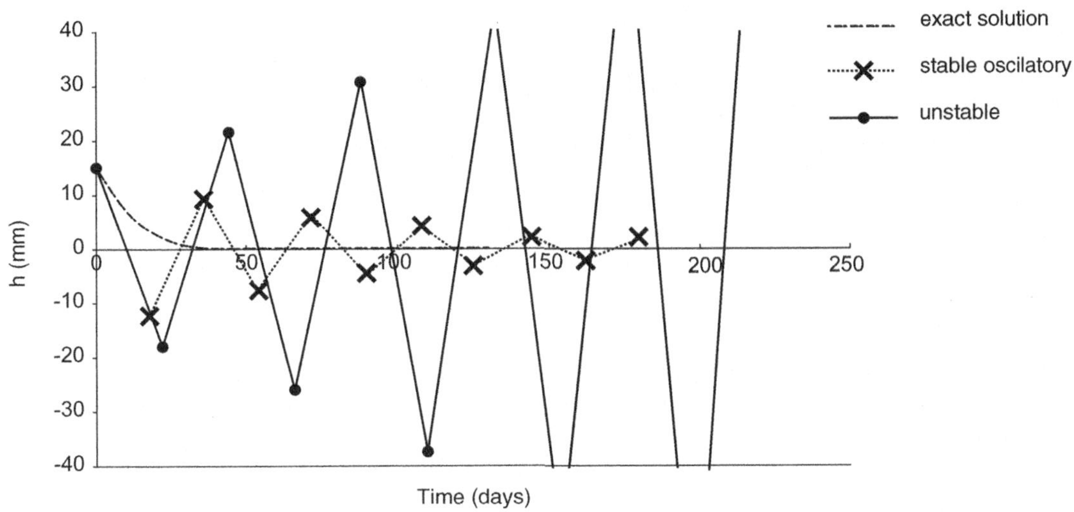

Figure 6.10 Exact and explicit solutions for the reservoir equation.

Relation of type (6.66) is analysed in behaviour as $\Delta t \to 0$. Because $\Delta t = t^n/n$, equation (6.66) can be written as:

$$h^n = \left[1 + \left(-\frac{\alpha t}{n}\right)\right]^n h^0 \tag{6.67}$$

Using the binomial expansion relation (6.67) is:

$$h^0 = h^0\left[1^n + 1^{n-1}\left[-\frac{\alpha t}{n}\right]\binom{n}{1} + 1^{n-2}\left[-\frac{\alpha t}{n}\right]^2\binom{n}{2} + \cdots\right] \tag{6.68}$$

where $\binom{n}{r}$ is the number of ways of choosing r items from a set of n items. Order of choosing is not important. The number of possible combinations is:

$$\binom{n}{r} = \frac{n!}{(n-r)!r!} \tag{6.69}$$

As $n \to \infty$, which is equivalent with $\Delta t \to 0$, n becomes so big that $n - 1 \approx n$, $n - 2 \approx n$, and so on. As a result:

$$\frac{n!}{(n-r)!r!} = n(n-1)(n-2)\cdots \to n^r \quad and \quad \binom{n}{r} \to \frac{n^r}{n!} \tag{6.70}$$

Hence:

$$h^n \to h^0\left[1^n + \frac{n}{1!}\left[-\frac{\alpha t}{n}\right] + 1^{n-2}\frac{n^2}{2!}\left[-\frac{\alpha t}{n}\right]^2 + \cdots\right] = \underbrace{h^0\left[1^n + \left[-\frac{\alpha t}{1}\right] + \frac{1}{2!}\left[-\frac{\alpha t}{1}\right]^2 + \cdots\right]}_{\text{Taylor series for } e^{-\alpha t}} \tag{6.71}$$

Hence:

$$h^n \to h^0 e^{-\alpha t} \tag{6.72}$$

which is the exact solution of the ODE, that is, equation (6.64).

The exact solution to the equation (6.63) *decays* with time. The value of h^n at a given time step always smaller than the value at the previous time step. The decay is *monotonic*, i.e. the sign of h^n does not change. Looking at equation (6.65), and taking into account the previous statement, if h^{n+1} is less then h^n yields:

$$0 \le 1 - \alpha\Delta t \le 1 \quad \leftrightarrow \quad 0 \le \Delta t \le \frac{1}{\alpha} \tag{6.73}$$

Equation (6.73) shows that the numerical solution decays *monotonically* only for a limited range of values of Δt, as it could be noticed from Figure 6.10.

Stability of numerical scheme (6.65) is analysed by applying the von Neumann principle:

$$\left|\frac{h^{n+1}}{h^n}\right| \le 1 \quad \leftrightarrow \quad |1 - \alpha\Delta t| \le 1 \tag{6.74}$$

For this to be true, the two following conditions must be verified simultaneously:

$$\begin{cases} \alpha\Delta t \le 2 \\ -\alpha\Delta t \le 0 \end{cases} \tag{6.75}$$

The first of the condition (6.74) is always verified because both the constant α and the time step Δt are positive. The second condition means that the time step cannot be larger than a maximum value defined by

$$\Delta t_{Max} = \frac{2}{\alpha} \tag{6.76}$$

Hence there is a *stability constraint* attached to the time step.

Consistency analysis aims to find out how accurately the *discretised* differential equation approximates the *real* differential equation. The analysis is done by using Taylor series expansions, expressing h^{n+1} as a function of h^n:

$$h^{n+1} = h^n + \Delta t\frac{dh}{dt} + \frac{\Delta t^2}{2}\frac{d^2h}{dt^2} + O(\Delta t^3) \tag{6.77}$$

The term $O(\Delta t^3)$ accounts for all the higher-order terms in the Taylor series expansion. Substituting the expression (6.77) into the numerical scheme (6.65), yields:

$$h^n + \Delta t \frac{dh}{dt} + \frac{\Delta t^2}{2}\frac{d^2h}{dt^2} + O(\Delta t^3) = (1 - k\Delta t)h^n \tag{6.78}$$

Simplifying h^n and dividing by Δt, yields:

$$\frac{dh}{dt} + \alpha h = -\frac{\Delta t}{2}\frac{d^2h}{dt^2} + O(\Delta t^3) \tag{6.79}$$

Comparing equation (6.79) with the original equation (6.63) shows that the discretised equation (6.79) is different from the original equation. The difference is the truncation error term, which becomes smaller when Δt decreases to zero. When Δt tends to zero, the term $O(\Delta t^2)$ decreases faster than the term $\Delta t d^2h/dt^2$ hence the term in Δt is the larger of the two. The truncation error is of first-order with respect to time step. The truncation error contains a second-order derivative d^2h/dt^2, which is not present in the real equation, consequently, the discretised equation approximates the real equation up to the first order only. The discretization is first-order accurate with respect to h.

6.4.2.2 *Implicit schemes*

The implicit scheme for the reservoir equation (6.63) is:

$$h^{n+1} = \frac{h^n}{1+\alpha\Delta t} \tag{6.80}$$

Stability analysis of the implicit scheme gives

$$\left|\frac{h^{n+1}}{h^n}\right| = \left|\frac{1}{1+\alpha\Delta t}\right| \le 1 \leftrightarrow |1+\alpha\Delta t| \ge 1 \tag{6.81}$$

Condition expressed by (6.81) is always verified. Since $\alpha > 0$, condition (6.81) is valid for all $\Delta t > 0$ and $\Delta t > 0$ by definition. The scheme always decays and never oscillates around 0. It is unconditionally stable, hence there is *no stability constraint* for the implicit method. (*Note*: because a scheme is stable, does not make it accurate.)

From physical point of view the reservoir becomes empty over time, hence h should decrease monotonically to 0, which is shown by the analytical solution. For the case of the explicit scheme Δh is invariably overestimated, to the extent that in some cases h is negative. In contrast, the implicit scheme, invariably underestimates the solution.

6.4.3 PDE: Convergence analysis for Preissmann scheme applied to advection equation

As seen there are a variety of numerical schemes that can be used to approximate the solution of the linear advection equation. One important scheme is the Preissmann scheme, also called the four point scheme or the box scheme. The four point scheme is often used for open channel flow modelling hence the study of its convergence is an important aspect to be looked at. Recall from chapter 4, section 4.5.2.4, using notations from Figure 4.18 and equations 4.56 for the scheme that problem to be solved is:

$$\begin{cases} \frac{\partial u}{\partial t} + a\frac{\partial u}{\partial x} = 0 \\ \left.\frac{\partial u}{\partial t}\right|_P = \psi\frac{u_{i-1}^{n+1} - u_{i-1}^n}{\Delta t} + (1-\psi)\frac{u_i^{n+1} - u_i^n}{\Delta t} \\ \left.\frac{\partial u}{\partial x}\right|_P = (1-\theta)\frac{u_i^n - u_{i-1}^n}{\Delta x} + \theta\frac{u_i^{n+1} - u_{i-1}^{n+1}}{\Delta x} \end{cases} \tag{6.82a,b,c}$$

The domain of computation of equation (6.82a) and the 4 point scheme is represented in Figures (4.19) and (4.18) of Chapter 4. The box scheme approximates the derivatives at a point P inside the discretization grid, using the values of the unknown function at grid points. There are in total M points of discretization. Advection velocity a is assumed constant and positive (i.e., $a > 0$), which implies that boundary condition is specified at $x = 0$, hence all values u_1^n are known at all time levels n of computation. The two weighting parameters θ and ψ ranges in the interval [0, 1].

Substitution of the derivatives discretization in advection equation yields:

$$\psi\frac{u_{i-1}^{n+1} - u_{i-1}^n}{\Delta t} + (1-\psi)\frac{u_i^{n+1} - u_i^n}{\Delta t} + a\left((1-\theta)\frac{u_i^n - u_{i-1}^n}{\Delta x} + \theta\frac{u_i^{n+1} - u_{i-1}^{n+1}}{\Delta x}\right) = 0, i = 2,\ldots, M \tag{6.83}$$

In each of the (M–1) equations of system (6.83) there is one unknown, hence each equation of the system can be written in a general form as:

$$u_i^{n+1} = \alpha \cdot u_{i-1}^n + \beta \cdot u_i^n + \gamma \cdot u_{i-1}^{n+1} \tag{6.84}$$

with values of the coefficients:

$$\alpha = \frac{\psi + (1-\theta)Cr}{1 - \psi + \theta \cdot Cr} \tag{6.84a}$$

$$\beta = \frac{1 - \psi - (1-\theta)Cr}{1 - \psi + \theta \cdot Cr} \tag{6.84b}$$

$$\gamma = \frac{-\psi + \theta \cdot Cr}{1 - \psi + \theta \cdot Cr} \tag{6.84c}$$

Courant number Cr is as previously defined:

$$Cr = \frac{a \cdot \Delta t}{\Delta x} \tag{6.85}$$

Based on the linear system of equations (6.83) all values of the unknown function $u(x, t)$ are computed at the time level $(n + 1)$ provided that values at time level n are already known. Computations should be carried out in a consecutive order starting from the left boundary and advancing towards the right boundary. It is not possible to reverse the order of computation. Because of such constraints (e.g., computation order) the scheme is an implicit one. Consider the example from Section 6.2.2, a 10 km long canal with an initial water level of 2 m and the input stage hydrograph as defined by relations (6.26). Computation of the solution is carried out over 10 hours. Results of solution for this problem, using numerical scheme (6.84) and different values for the weighting coefficients (θ and ψ), are shown in Figures 6.11, 6.12 and 6.13, at the discretization grid point $x = 5$ km.

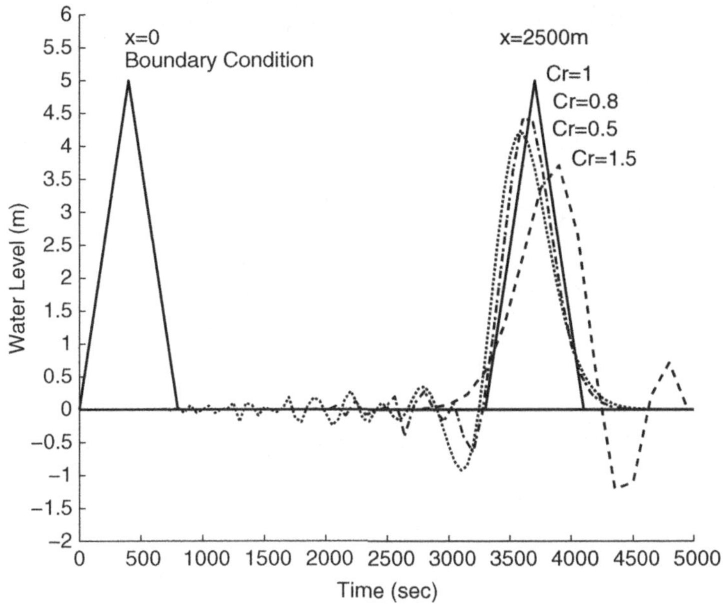

Figure 6.11 Solution of advection equation using the 4-point numerical scheme for $\theta = 0.5$ and $\psi = 0.5$.

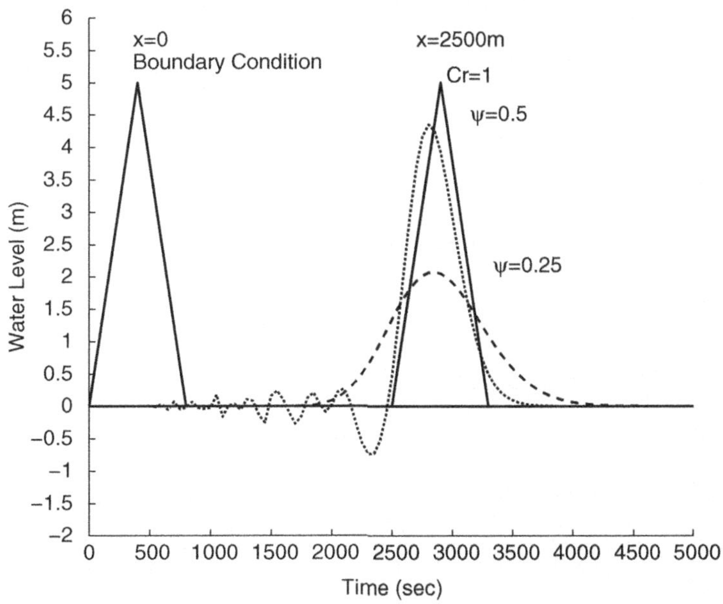

Figure 6.12 Solution of advection equation using the 4-point numerical scheme for $\theta = 0$ and $Cr = 0.5$.

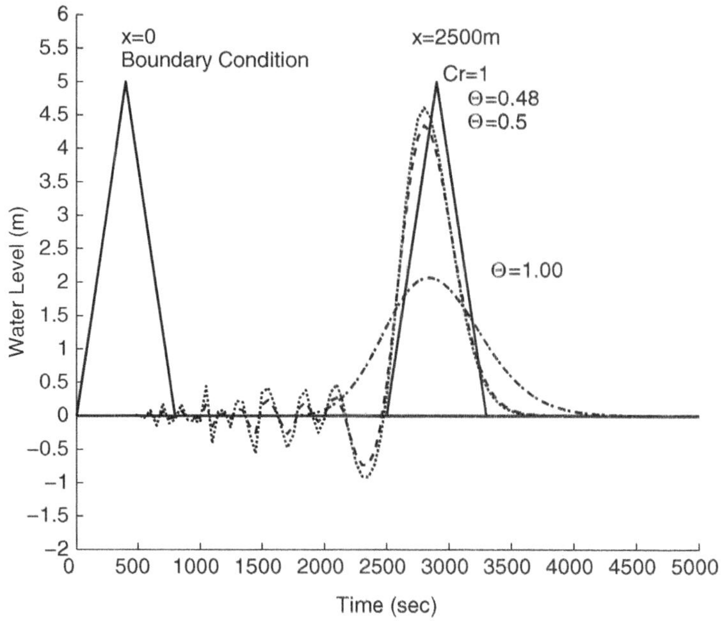

Figure 6.13 Solution of advection equation using the 4-point numerical scheme for $\psi = 0$ and $Cr = 0.5$.

Figures 6.11, 6.12, 6.13 shows that there are values of θ, ψ, Cr for which solutions are either oscillating or dampening. The physical phenomenon represented in the equation is just advection, hence the errors in solution are of numerical nature. Analysis of the error of the solution must give the constraints of application of the numerical scheme.

As shown by relation (6.33) the error of the numerical scheme solution satisfy the equation of the applied numerical scheme, hence in the case of box scheme the error is:

$$TE_j^{n+1} = \alpha \cdot TE_{j-1}^n + \beta \cdot ET_j^n + \gamma \cdot TE_{j-1}^{n+1} \tag{6.86}$$

(*Note*: Because Fourier series are used further for the analysis of the error, indices j are used for space discretization points, in order to avoid confusion with the imaginary unit i.)

Using Fourier series to represent the solution error yields:

$$A^{n+1} \cdot e^{i \cdot \varphi \cdot j} = \alpha \cdot A^n \cdot e^{i \cdot \varphi (j-1)} + \beta \cdot A^n \cdot e^{i \cdot \varphi \cdot i} + \gamma \cdot A^{n+1} \cdot e^{i \cdot \varphi (j-1)} \tag{6.87}$$

From equation (6.87) the amplification factor is computed as:

$$\frac{A^{n+1}}{A^n} = A = \frac{\beta + \alpha \cdot e^{-i \cdot \varphi}}{1 - \gamma \cdot e^{-i \cdot \varphi}} \tag{6.88}$$

The amplitude of the error should not increase, it is required that $|A| < 1$, hence:

$$A = \frac{\beta + \alpha \cdot \cos\varphi - i \cdot \alpha \cdot \sin\varphi}{1 - \gamma \cdot \cos\varphi + i \cdot \gamma \cdot \sin\varphi} \tag{6.89}$$

Rearranging and using properties of complex numbers, one obtains:

$$A = \frac{\beta + (\alpha - \beta \cdot \gamma)\cos\varphi - \alpha \cdot \gamma}{1 - 2\gamma \cdot \cos\varphi + \gamma^2} - \frac{(\alpha + \beta \cdot \gamma)\sin\varphi}{1 - 2\gamma \cdot \cos\varphi + \gamma^2} i \tag{6.90}$$

Stability condition yields:

$$|A| = \sqrt{\left[\frac{\beta + (\alpha - \beta \cdot \gamma)\cos\varphi - \alpha \cdot \gamma}{1 - 2\gamma \cdot \cos\varphi + \gamma^2}\right]^2 + \left[\frac{(\alpha + \beta \cdot \gamma)\sin\varphi}{1 - 2\gamma \cdot \cos\varphi + \gamma^2}\right]^2} \leq 1 \tag{6.91}$$

In inequality (6.91) $\varphi \in [0, \pi]$, hence the inequality is checked for the minimum and maximum values of φ.

If = π, inequality (6.91) is:

$$\left(\frac{\beta - \alpha}{1 + \gamma}\right)^2 \leq 1 \tag{6.92}$$

equivalent with

$$-1 \leq \frac{\beta - \alpha}{1 + \gamma} \leq 1 \tag{6.93}$$

Replacing the expression of coefficients α, β, and γ, yields:

$$-1 \leq \frac{1 - 2\psi - 2(1 - \theta)Cr}{1 - 2\psi + 2\theta \cdot Cr} \leq 1 \Leftrightarrow \begin{cases} -1 \leq \dfrac{2Cr}{1 - 2\psi + 2\theta \cdot Cr} \\ 1 - \dfrac{2Cr}{1 - 2\psi + 2\theta \cdot Cr} \leq 1 \end{cases} \tag{6.94}$$

If = 0, inequality (6.91) is:

$$\left|\frac{\beta + \alpha}{1 - \gamma}\right| \leq 1 \tag{6.95}$$

equivalent with

$$-1 \leq \frac{\beta + \alpha}{1 - \gamma} \leq 1 \tag{6.96}$$

Replacing the expression of coefficients α, β, and γ, yields:

$$-1 \leq \frac{1 + 2\psi}{1 - 2\psi + \theta \cdot Cr} \leq 1 \Leftrightarrow \begin{cases} 0 \leq \dfrac{2 + \theta \cdot Cr}{1 - 2\psi + \theta \cdot Cr} \\ \dfrac{4\psi + \theta \cdot Cr}{1 - 2\psi + 2\theta \cdot Cr} \leq 0 \end{cases} \tag{6.97}$$

Relations (6.94) and (6.97) are satisfied for:

$$\theta \geq \frac{1}{2} \quad \text{and} \quad \psi \leq \frac{1}{2} \tag{6.98}$$

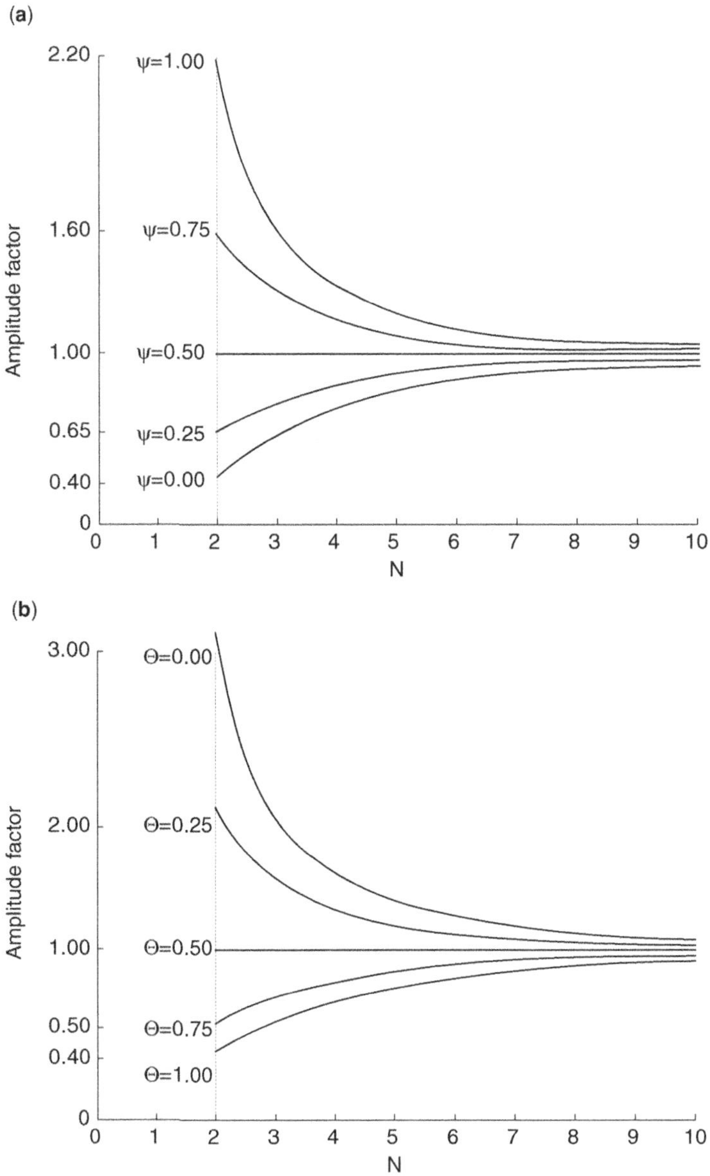

Figure 6.14 Amplification factor versus N for advection equation for $Cr = 2$: (a) different θ values and $\psi = 0.5$; (b) different ψ values and $\theta = 0.5$.

Condition (6.98) ensures the denominator in relations (6.94) and (6.98) is not zero. Moreover the inequalities are satisfied for any value of Courant number, *Cr*, hence the scheme is unconditionally stable if (6.98) is valid.

In the case that values of parameters θ and ψ are zero the explicit FTBS scheme is obtained, which has different numerical properties than the box scheme, as shown earlier in the chapter.

Further it is important to analyse the conditions for which box scheme does not introduce artificial dissipation or dispersion, because the analytical solution of the advection equation is a wave that has a specific amplitude and celerity which should not be changed by the numerical solution. The analysis is carried out, by examining the amplitude and phase errors. The amplitude errors are given by the modulus of the amplification factor, which for box scheme is:

$$|A| = \left\{ \left[\frac{\beta + (\alpha - \beta \cdot \gamma)\cos\left(\frac{2\pi}{N}\right) - \alpha \cdot \gamma}{1 - 2\gamma \cdot \cos\left(\frac{2\pi}{N}\right) + \gamma^2} \right]^2 + \left[\frac{(\alpha + \beta \cdot \gamma)\sin\left(\frac{2\pi}{N}\right)}{1 - 2\gamma \cdot \cos\left(\frac{2\pi}{N}\right) + \gamma^2} \right]^2 \right\}^{1/2} \tag{6.99}$$

Recall that N is the ratio between the wave length and the size of the space discretization (Δx).

Representing graphically the function A given by (6.99) for different values of Courant number (see Figure 6.14), it can be seen that for $Cr = 1$, and $\theta = \psi = 0.5$ the box scheme is free from dissipation, while for, $\theta \geq 0.5$ and $\psi \geq 0.5$, the scheme increases the wave amplitude over time, which leads to instabilities. Small values of N and large Courant numbers creates damping, hence the number of discretization points over the wave length should be carefully selected.

6.4.4 PDE: Convergence analysis for diffusion equation

Analysis of convergence for diffusion equation using a FTCS scheme is analysed. The equations are:

$$\begin{cases} \dfrac{\partial u}{\partial t} - D\dfrac{\partial^2 u}{\partial x^2} = 0 \\ \dfrac{\partial u}{\partial t} = \dfrac{u_i^{n+1} - u_i^n}{\Delta t} \\ \dfrac{\partial u}{\partial x} = D\dfrac{u_{i+1}^n - 2u_i^n + u_{i-1}^n}{\Delta x} \end{cases} \tag{6.100a, b, c}$$

where D is the diffusion coefficient. Substitution of the derivatives discretization in advection equation yields:

$$u_i^{n+1} = u_i^n + r \cdot (u_{i+1}^n - 2u_i^n + u_{i-1}^n), \quad with\ r = \frac{D\Delta t}{(\Delta x)^2} \tag{6.101}$$

The error of the numerical scheme solution satisfy the equation of the applied numerical scheme, hence using Fourier series to represent the solution error yields:

$$A^{n+1} \cdot e^{i \cdot \varphi \cdot j} = A^n(1 + r \cdot (e^{i \cdot \varphi \cdot j} + e^{-i \cdot \varphi \cdot i} - 2)) \tag{6.102}$$

(*Note*: Because Fourier series are used further for the analysis of the error, indices j are used for space discretization points, in order to avoid confusion with the imaginary unit *i*.)

From equation (6.102) the amplification factor is computed as:

$$\frac{A^{n+1}}{A^n} = A = 1 - 2r \cdot (1 - \cos\varphi) \tag{6.103}$$

The amplitude of the error should not increase, it is required that $|A| < 1$, hence:

$$r \leq 1/2 \tag{6.104}$$

which shows that the scheme is conditionally stable.

Chapter 7

River system modelling and flood propagation

7.1 INTRODUCTION

Rivers are very important for mankind because they provide a source for drinking water and food production. Moreover they are important as a habitat for wildlife and for maintaining the equilibrium of the ecosystems formed along them. Apart from these natural benefits, rivers can also be used as means of transportation of goods and as recreation areas. Though sometimes rivers are causing floods, human settlements were always located in the river floodplains because of the social and economic benefits that the rivers are bringing (Knight, 2006; Novak, 1990). As societies developed their need for managing the river systems along which they live grew and changes started to be implemented in order to take advantage of the river's water availability or to protect the settlements against flood hazards. Communities have been developing and managing water resources in river basins in order to satisfy their demands for water quantity and quality or to cope with extreme events such as floods and droughts, by building dams to have access to water, irrigation channels to take water from rivers, dykes to protect the floodplains from flood risks, and so on. These developments were challenges to hydraulic engineers who needed to understand and evaluate the mechanism of flow in a river as well as the mechanism of flood formation. Stream flow predictions at particular locations in a river channel were done by engineers since a long time, and lately due to the advent of computers their tasks is facilitated by the availability of river system modelling tools, that are able to help in long and short term planning. The concept of River Basin Management (RBM) emerged and is accepted as a central element in all policy directives regarding rivers worldwide. The concept emerged as a holistic approach to river systems management and tries to eliminate the negative effect of the competitive interventions in a system due to the different water resources needs, such as unbalanced water resources exploitation between upstream and downstream of a river system.

In the last four decades model codes have been developed for water resources and environmental analyses. These codes have been developed by academia, governmental and consultancy agencies and they vary in complexity from simple conceptual codes to complex ones based on physical processes. Majority of these codes are nowadays embedded in a tool, with complex graphical user interfaces. Further more, many of these tools are embedded in Decision Support Systems (DSSs), that are not only a framework for providing data and information to decision makers, but they are also platforms for sharing and disseminating information to the public (Popescu *et al.*, 2012).

The DSS systems applied in RBM are using river system modelling, because it allows for understanding the behaviour of the river system as a response to some input variables such as discharges in an upstream part and/or conditions at the downstream boundary of a river. Moreover water quality analysis can be carried out if the hydrodynamic of flow in a river is known. A model of a river can help the decision making process by equipping the decision makers with a tool to objectively evaluate the effect of certain decisions. Mathematical (hydraulic) models for rivers, channel and pipe systems, emerged in 1980 and are the most widely used codes. They are used for predicting flows, travel times, and water level variations in rivers, channels and canal systems.

Floodplains along the rivers are also important because they are the preferred location for living, however the effect of floods are loss of life, loss of property and adverse impacts on economic activities. Hence there is the need for accurate and reliable flood analysis and management of flood from both forecasting and mitigating points of view. Hydraulic models of river flow simulations became important instruments to address these issues.

Development of river models started during 1970s. River models are in general part of river basin models, which were developed especially for planning purposes (Zagona *et al.*, 2001; Jonoski & Popescu, 2004), both long and short-term. River basin models are used for water management purposes using different approaches and technologies. There are a number of models able to represent different hydrological processes, as well as river hydrodynamics. Examples of such systems that include sophisticated graphical user interfaces are MIKE SHE (DHI, 2012), HEC suite of models (USACE models) and SOBEK (Deltares, 2010). There are many other examples of such systems. A comprehensive overview of existing river modelling systems is given by Horrit and Bates (2002) .

7.2 RIVER SYSTEMS MODELLING

Computational Hydraulics is a valuable tool for designer engineers because it allows for prediction of flow in a particular river and under particular boundary and initial conditions without making measurements in the field. The application of Computational Hydraulics to river related issues covers a wide range of problems such as flood alleviation, river rehabilitation (water quality and quantity), operation of reservoirs, and morphological predictions.

The equations that govern the motion of flow can be solved analytically only for particular very simple problems, therefore the use of numerical methods for solving these equations is a good way forward. Rivers are having irregular topographies, which are leading to mixed regions of flows. This poses challenges to the numerical solutions of the flow equations, because they do need to solve these problems under a wide range of boundary and initial conditions. Moreover flow around and through structures (weirs, gates, etc.) needs to be addressed as well. All numerical approaches are subject to stability constraints, which restricts the selection of the time and space step and can extend the computational time.

A number of numerical approaches exist to solve the flow equation processes. Corresponding to the approaches presented a number of mathematical models exists for solving flow in rivers. These are ranging from 1D Saint-Venant to 3D Navier-Stokes equations. All equations are based on the same physical principles and the selection of one model for a particular problem depends on the actual problem considered and the requirements on the solution. This chapter shows the numerical solution approach for the one dimensional Saint-Venant equations, which are simple as compared with the complex Navier-Stokes equations, however they are able to predict enough useful information for river engineers.

Until recently the non-conservative form of the shallow water equations was used to compute flow patters, however this is not useful for problems that involve rapidly changing transitional flow, for which sharp fronts and rapid changes in velocity occur. In this section of the book the conservative equations of flow are considered and solutions are shown. All demonstrations are made on rectangular grids because historically they have remained a popular choice and are the basis for a number of widely used algorithms (Sturm, 2001).

For channels of irregular topographies, the flow can be described by the Saint-Venant equations, as shown in Chapter 2. The Saint-Venant equations for one-dimensional flow are:

- continuity equation (CE):

$$\frac{\partial Q}{\partial x} + b\frac{\partial h}{\partial t} = q \quad or \quad \frac{1}{b}\cdot\frac{\partial Q}{\partial x} + \frac{\partial h}{\partial t} = \frac{q}{b} \tag{7.1}$$

- momentum equation (ME):

$$\frac{\partial Q}{\partial t} + \frac{\partial}{\partial x}\left(\beta\frac{Q^2}{A}\right) + gA\frac{\partial h}{\partial x} - gAS_O + gA\frac{|Q|Q}{K^2} = 0 \tag{7.2}$$

where Q is the discharge (m³/s); b the storage width at the water surface level (m); h water depth (m); A the cross-sectional area (m²); q the lateral discharge (m³/s); K the conveyance (m³/s); β the Boussinesq coefficient (–); g gravity (m/s²) and S_o the bed slope of the channel (–).

The system of equation formed by (7.1) and (7.2) has an analytical solution just for very few particular cases. These are usually solved using numerical approximation. There are many approaches to solve the system of equations however two of them, based on FDM, are the most used ones in practice, and are detailed further, the four point Preissmann scheme and the six-point Abbott Ionescu scheme.

Prior to building any mathematical model and performing computations for a specific river, the geometry of the river needs to be known (Figure 7.1), by determining the course of it (the river bottom in a number of cross-sections. River systems are schematised into a simplified river network using a node-link structure (see Figure 7.1). The river network begins and ends with a node, and all nodes are interconnected. From the map representation of the river network, the extent of the model is selected by positioning the upstream and downstream boundaries. A computational space step Δx (usually constant) is selected as well. The obtained nodes represent the computational points on the grid and all relationship between the water depth h in these nodes and a number of water depth related parameters, such as width of water table (B), or cross-sectional area (A) are computed in all discretization points. Computational nodes do not usually correspond to the measured cross-sections and interpolation is used to determine values of parameters in computational nodes.

7.2.1 Preissmann solution

The Preissmann scheme for solving one-dimensional Saint-Venant equation was first introduced in 1961 (Preissmann, 1961) in 'Propagation of translatory waves in channels and rivers'. Preissmann scheme, also known as the four-point weighted implicit or box scheme is the most popular scheme used. Popular commercial software codes for solving river related problems, based on one dimensional Saint-Venant equations, such as HEC-RAS (Brunner, 2010) and DUFLOW (Clemmens *et al.*, 1993) are using Preissmann scheme for solution of river flows. Preissmann scheme is suitable for relatively flat bed slopes. Dry-bed conditions are not allowed to occur during simulation in the codes using Preissmann scheme (Ogden & julien 2002). The scheme is applicable for both subcritical and supercritical flow, however the use of the scheme should not be extended to trans-critical flow regime (Chau, 1990; Meselhe & Holly, 1997). The limitation is due to the requirements on boundary conditions for different flow regimes. Abbott *et al.* (1991) shows that this issue can be solved by minimizing the effect of the convective acceleration terms in momentum equations in order to ensure that flow remains in sub-critical regime. Hence the convective acceleration term is multiplied by a factor (Djordevic *et al.*, 2004) which ensures that flow is always subcritical.

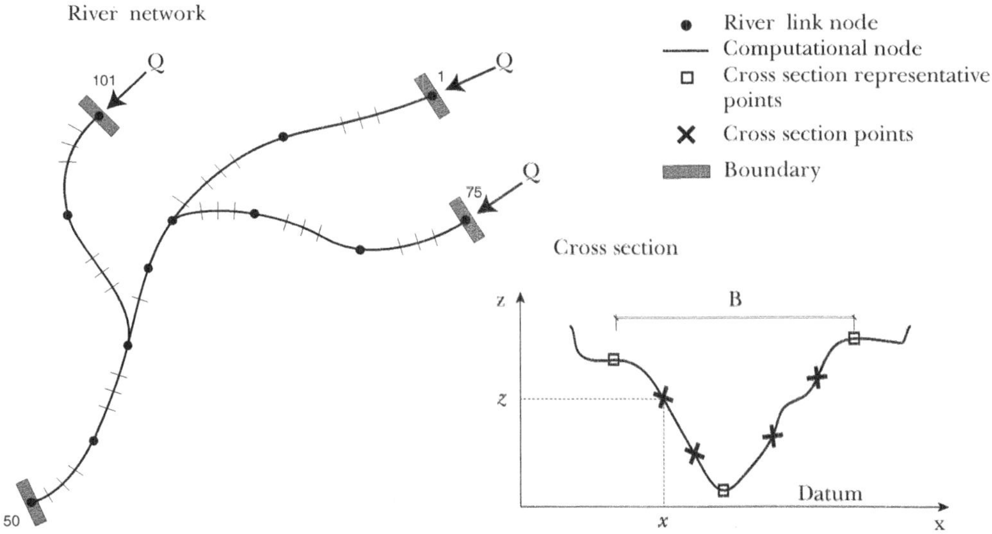

Figure 7.1 River network, cross-sections and computational nodes.

Discretization of the equations (7.1) and (7.2) is done on a computational domain (x, t), as it is represented in Figure 7.2. using the stencil from Figure 4.18 of Chapter 4 of the book.

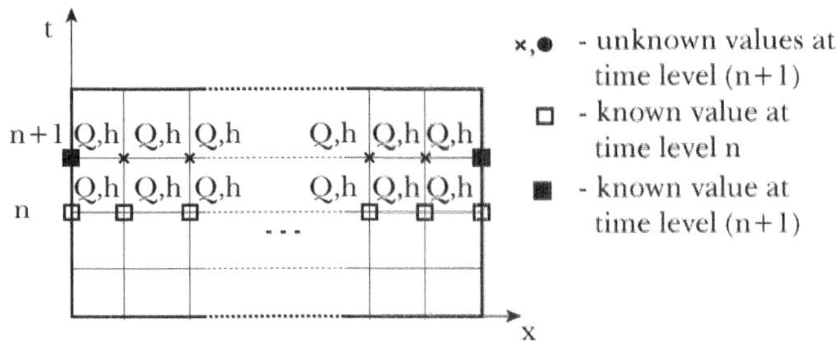

Figure 7.2 Computational domain for the Saint-Venant equations for Preissmann scheme solution.

The discretization grid considered is a structured one that computes values of Q and h in all points of the computational domain. Solution of the system of equations defined by (7.1) is solved for $x \in [0, L]$ and $t \geq 0$. Applying the Preissmann scheme to the terms in equation (7.1) yields:

$$\frac{\partial Q}{\partial x} = \theta \left.\frac{\partial Q}{\partial x}\right|_{n+1} + (1-\theta)\left.\frac{\partial Q}{\partial x}\right|_{n} = \theta \frac{Q_{i+1}^{n+1} - Q_i^{n+1}}{\Delta x} + (1-\theta)\frac{Q_{i+1}^{n} - Q_i^{n}}{\Delta x} \tag{7.3}$$

$$\frac{\partial h}{\partial t} = \psi \left.\frac{\partial h}{\partial t}\right|_{i+1} + (1-\psi)\left.\frac{\partial h}{\partial t}\right|_{i} = \psi \frac{h_{i+1}^{n+1} - h_{i+1}^{n}}{\Delta t} + (1-\psi)\frac{h_i^{n+1} - h_i^{n}}{\Delta t} \tag{7.4}$$

Substituting (7.3) and (7.4) in equation (7.1) and grouping the unknown variables, Q_i^{n+1}, h_i^{n+1}, on the left hand side results in:

$$A1_i Q_i^{n+1} + B1_i h_i^{n+1} + C1_i Q_{i+1}^{n+1} + D1_i h_{i+1}^{n+1} = E1_i \tag{7.5}$$

with values of the coefficients, in case $q = 0$, given by:

$$A1_{i+1} = \frac{-\theta}{\Delta x} \tag{7.6a}$$

$$B1_{i+1} = (1-\psi)\frac{b}{\Delta t} \tag{7.6b}$$

$$C1_{i+1} = \frac{\theta}{\Delta x} \tag{7.6c}$$

$$D1_{i+1} = \frac{b\psi}{\Delta t} \tag{7.6d}$$

$$E1_{i+1} = \frac{1-\theta}{\Delta x}\left(Q_j^n - Q_{j+1}^n\right) + \frac{b}{\Delta t}[(1-\psi)h_j^n + \psi h_{j+1}^n] \tag{7.6e}$$

Applying the same approach (substituting (7.3) and (7.4) in equation (7.2)) to momentum equation, yields:

– Discretization of first term of momentum equation:

$$\frac{\partial Q}{\partial t} = \psi \frac{Q_{i+1}^{n+1} - Qh_{i+1}^{n}}{\Delta t} + (1-\psi)\frac{Q_i^{n+1} - Q_i^{n}}{\Delta t} \tag{7.7a}$$

– Discretization of second term of momentum equation:

$$\frac{\partial}{\partial x}\left(\beta \frac{Q^2}{A}\right) = \frac{\beta}{\Delta x}\left[\left.\left(\frac{Q^2}{A}\right)\right|_{i+1} - \left.\left(\frac{Q^2}{A}\right)\right|_{i}\right]$$

Considering the following linearization for the non linear term Q^2 (Abbot & Basco, 1989):

$$Q^2 = Q^n \times Q^{n+1}$$

and

$$A_{i+1} = A_{i+1}^{n+1/2} = \frac{A_{i+1}^{n} + A_{i+1}^{n+1}}{2}$$

the second term of the momentum equation is:

$$\frac{\partial}{\partial x}\left(\beta\frac{Q^2}{A}\right) = \beta\left[\frac{Q_{i+1}^{n}\times Q_{i+1}^{n+1}}{\Delta x\left(A_{i+1}^{n+1/2}\right)} - \frac{Q_{i}^{n}\times Q_{i}^{n+1}}{\Delta x\left(A_{i}^{n+1/2}\right)}\right] \tag{7.7b}$$

– Discretization of third term of momentum equation:

$$\begin{aligned} gA\frac{\partial h}{\partial x} &= gA_{i+1/2}^{n+1/2}\left[\theta\left.\frac{\partial h}{\partial x}\right|_{n+1} + (1-\theta)\left.\frac{\partial h}{\partial x}\right|_{n}\right] \\ &= gA_{i+1/2}^{n+1/2}\left[\theta\frac{h_{i+1}^{n+1}-h_{i}^{n+1}}{\Delta x} + (1-\theta)\frac{h_{i+1}^{n}-h_{i}^{n}}{\Delta x}\right] \end{aligned} \tag{7.7c}$$

– Discretization of fourth term of momentum equation:

$$gAS_O = gA_{I+1/2}^{N+1/2}S_O \tag{7.7d}$$

– Discretization of fifth term of momentum equation is done using the coefficient ψ because Q and K are varying over time:

$$gA\frac{|Q|Q}{K^2} = g\psi A\left.\frac{|Q|Q}{K^2}\right|_{i+1} + g(1-\psi)A\left.\frac{|Q|Q}{K^2}\right|_{i}$$

hence:

$$gA\frac{|Q|Q}{K^2} = g\left[\psi A_{i+1}^{n+\frac{1}{2}}\frac{\left|Q_{i+1}^{n}\right|Q_{i+1}^{n+1}}{\left(K_{i+1}^{n+\frac{1}{2}}\right)} + (1-\psi)A_{i}^{n+\frac{1}{2}}\frac{\left|Q_{i}^{n}\right|Q_{i}^{n+1}}{\left(K_{i}^{n+\frac{1}{2}}\right)^2}\right] \tag{7.7e}$$

Grouping the unknown variables, Q_i^{n+1}, h_i^{n+1}, on the left hand side results in:

$$A2_iQ_i^{n+1} + B2_ih_i^{n+1} + C2_iQ_{i+1}^{n+1} + D2_ih_{i+1}^{n+1} = E2_i \tag{7.8}$$

with coefficients:

$$A2_{i+1} = \left[\frac{1-\psi}{\Delta t} - \frac{\beta Q_j^n}{\Delta x\left(A_i^{n+1/2}\right)} + (1-\psi)A_i^{n+1/2}\frac{\left|Q_i^n\right|}{\left(K_i^{n+1/2}\right)^2}\right] \tag{7.9a}$$

$$B2_{i+1} = \left[-gA_{i+1/2}^{n+1/2}\frac{\theta}{\Delta x}\right] \tag{7.9b}$$

$$C2_{i+1} = \left[\frac{\psi}{\Delta t} + \beta \frac{Q_{i+1}^n}{\Delta x (A_{i+1}^{n+1/2})} + g\psi A_{i+1}^{n+1/2} \frac{\left|Q_{i+1}^n\right|}{\left(K_{i+1}^{n+1/2}\right)^2} \right] \tag{7.9c}$$

$$D2_{i+1} = \left[g\psi A_{i+1}^{n+\frac{1}{2}} \frac{\theta}{\Delta x} \right] \tag{7.9d}$$

$$E2_{i+1} = \frac{1-\psi}{\Delta t} Q_i^n + \frac{\psi}{\Delta t} Q_{i+1}^n + gA_{i+1/2}^{n+1/2} \left[\frac{1-\theta}{\Delta x} \right] \left(h_{i+1}^n - h_i^n\right) + gA_{i+1/2}^{n+1/2} So \tag{7.9e}$$

The pair of equations (7.5) and (7.8) are written for each point of the discretization grid, that is for $i = 1, 2 \ldots, M-1$, to obtain a system of $2(M-1)$ equations that has $2M$ unknowns. The system is completed with two more relations at the boundary of the domain:

- for node $i = 1$:

$$bc(0) \cdot h_1^{n+1} + (1 - bc(0)) \cdot Q_1^{n+1} = bc(0) \cdot h_1(t^{n+1}) + (1 - bc(0)) \cdot Q_1(t^{n+1}) \tag{7.10a}$$

- for node $i = M$:

$$\begin{aligned} &bc(L) \cdot h_M^{n+1} + (1 - bc(L)) \cdot Q_M^{n+1} \\ &\quad = bc(L) \cdot h_M(t^{n+1}) + (1 - bc(L)) \cdot Q_M(t^{n+1}) \end{aligned} \tag{7.10b}$$

In equations (7.10) $bc(x)$ is a coefficient which takes the value 0 or 1 depending on the type of the specified boundary condition, that is coefficient value 1 means water depth is specified as boundary condition and value zero means a discharge boundary condition.

A number of $2M$ equations are obtained for a number of $2M$ unknowns. Such a system can be solved using matrix calculations or in an iterative method by using the so-called Double Sweep Algorithm.

Solution of the linear system of equation is given by the solution of the bandwidth matrix:

$$\begin{bmatrix} \text{Left BC} & & & & & & & & & & \\ A1_1 & B1_1 & C1_1 & D1_1 & & & \cdots & & & & \\ A2_1 & B2_1 & C2_1 & D2_1 & & & \cdots & & & & \\ & & A1_2 & B1_2 & C1_2 & D1_2 & \cdots & & & & \\ & & A2_2 & B2_2 & C2_2 & D2_2 & \cdots & & & & \\ \cdot & & & & & & \cdots & & & & \cdot \\ \cdot & & & & & & \cdots & & & & \cdot \\ \cdot & & & & & & \cdots & & & & \cdot \\ & & & & & & \cdots & A1_{M-1} & B1_{M-1} & C1_{M-1} & D1_{M-1} \\ & & & & & & \cdots & A2_{M-1} & B2_{M-1} & C2_{M-1} & D2_{M-1} \\ & & & & & & \cdots & & & & \text{Right BC} \end{bmatrix} \begin{bmatrix} Q_1 \\ h_1 \\ Q_2 \\ h_2 \\ \\ \cdot \\ \cdot \\ \cdot \\ \\ Q_M \\ h_M \end{bmatrix} = \begin{bmatrix} \text{Value from BC} \\ E1_1 \\ E2_1 \\ E1_2 \\ E2_2 \\ \cdot \\ \cdot \\ \cdot \\ E1_{M-1} \\ E2_{M-1} \\ \text{Value from BC} \end{bmatrix}$$

An alternative solution to matric calculations is the one given by the Double Sweep Algorithm, which takes into account that at each point of discretization grid a relationship can be written between the unknowns, Q_i^{n+1} and h_i^{n+1}, as well as between , Q_{i+1}^{n+1}, h_i^{n+1}, in the form:

$$Q_i^{n+1} = F_i h_i^{n+1} + G_i \tag{7.11a}$$

$$h_i^{n+1} = H_j Q_{i+1}^{n+1} + I_i h_{i+1}^{n+1} + J_i \tag{7.11b}$$

Coefficients *H, I, J, F, G* are derived by replacing relations (7.11) into the system of equations formed by equations (7.5) and (7.8):

$$H_i = \frac{-C1_i}{A1_i F_i + B1_i} \tag{7.12a}$$

$$I_i = \frac{-D1_i}{A1_i F_i + B1_i} \tag{7.12b}$$

$$J_i = \frac{E1_i - A1_i G_i}{A1_i F_j + B1_i} \tag{7.12c}$$

$$F_{i+1} = \frac{D2_i(A1_i F_i + B1_i) - D1_i(A2_i F_i + B2_i)}{C1_j(A2_j F_j + B2_j) - C1_j(A1_j F_j + B1_j)} = \frac{-I_i(A2_i F_i + B2_i) - D2_i}{H_i(A2_i F_i + B2_i) + C2_i} \tag{7.12d}$$

$$G_{i+1} = \frac{E2_i - A2_i F_i J_i - A2_i G_i - B2_i J_i}{H_i(A2_i F_i + B2_i) + C2_i} \tag{7.12e}$$

The step by step procedure to solve the system of equations for time level $(n + 1)$ using the double sweep algorithm is illustrated in Figure 7.3 and follows the steps:

(1) Initial data in terms of water level and discharge are specified for all grid points from *point* 1 to *point M* of discretization grid, at time level *n*.
(2) In order to start the computation of *Q* and *h* at time level $(n + 1)$ the assumption that all values of water level and discharge at time level $(n + 1)$ are the same as the values of water level and discharge at time level *n* is made.
(3) Computation of all $A1_i, B1_i, C1_i, D1_i, E1_i$, $A2_i, B2_i, C2_i, D2_i, E2_i$ coefficients at level $(n + 1)$ is done using formulas (7.8*a–e*). Coefficients are computed based on available data at time level *n* and $(n + 1)$.
(4) Boundary data at the upstream end is provided in terms of discharge or water level in grid discretization point 1. Coefficient *bc*(0) (see equation (7.11*a*)) is set to the appropriate choice.
(5) Values of *Q* and *h* at time level $(n + 1)$, at the left boundary (upstream) are used to compute coefficients *F* and *G* in point 1 of the discretization grid (i.e., F_1 and G_1).
(6) *Forward sweep*: Using recurrence relations (7.12*d–e*), values of coefficients F_i and G_i are computed by sweeping across the grid (from point 1 to point *M*). Coefficients H_i, I_i, and J_i are calculated using relations (7.12*a–c*). All values are computed at discretization points of the grid, as opposed to coefficients *A,B,C,D,E*, which are computed in grid cells (see Figure 7.3).
(7) Based on downstream boundary condition, which may be either discharge or water level the other unknown (water level or discharge, respectively) is computed using relations (7.11*a*).

(8) *Backward sweep*: Using previously computed coefficients F_i, G_i, H_i, I_i and J_i, discharge and water depth at time level $(n + 1)$ are calculated by sweeping back across the grid (from point M to point 1) calculating discharge and water depth at time level $(n + 1)$ using relations (7.12*a–b*). Discharge and water depth are determined at time level $(n + 1)$ for all points.

(9) Computed values for water level and discharge, obtained in step 8, are compared with the values from previous iteration. If they are not in accordance with a comparison criterion step 10 is applied.

(10) *Iterate*: Computation is repeated starting at step 3, using the values obtained in step 8 to re-compute all the coefficients in a forward sweep.

(11) Iterations are repeated until a certain condition is fulfilled (difference between two computed values, maximum number of iterations, etc.).

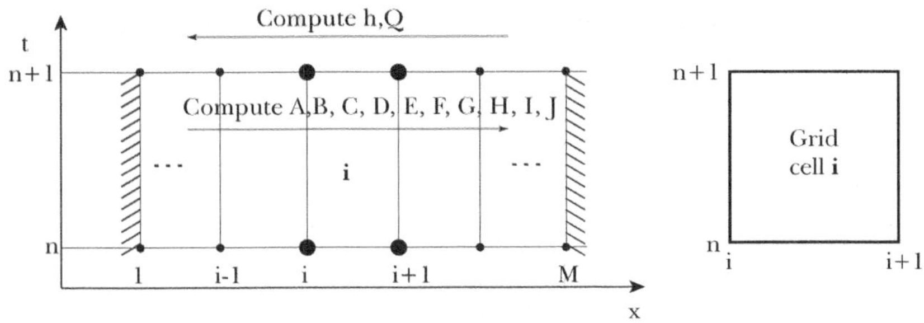

Figure 7.3 Double sweep algorithm.

7.2.2 Abbott-Ionescu solution

A second FDM approach for modelling of unsteady river flow is the Abbot-Ionescu scheme, which is implemented by Danish Hydraulic Institute in their MIKE 11 river flow model component of Mike Zero.

The scheme is based, as shown in Chapter 2, on a staggered grid (alternating grid-points). Based on this approach the solution is sought as Q and h values at alternate points of the discretization grid (see Figure 7.4).

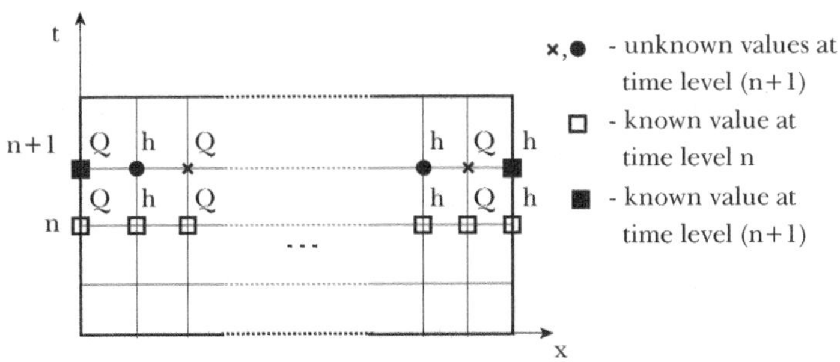

Figure 7.4 Computational domain for the Saint-Venant equations for Abbott-Ionescu scheme solution.

Figure 7.4. shows the one dimensional (x, t) computational domain for which the method is applied to solve the Saint-Venant system of equations. The Abbot-Ionescu scheme has been developed to solve subcritical river flow, hence one boundary condition has to be defined upstream and one downstream. For showing the principle of the method the boundary condition upstream has been selected to be inflow (the most common condition) and the downstream one has been selected to be water depth, which means that the time series of inflow $Q(x = 0, t)$ and $h(M \cdot \Delta x, t)$ are known before the computation of the solution, in the entire domain, starts to be determined. The number of discretization nodes along the x axis is M starting with 1 as the first point. An odd number of points are required for the discretization.

The staggered grid definition and the boundary conditions imposes that at all time levels, in all even points of the discretization grid ($1 = 0, 2, 4\ldots, M - l$) a Q value is computed; and in all odd ones ($i = 1, 3, 5, \ldots, M$), a h value is computed.

In order to demonstrate the method it is considered that all values of discharge and water depth, at time level n, Q^n and h^n, respectively, are known; and they are the basis for computing the values at time level $(n + 1)$ where Q^{n+1} and h^{n+1} are not yet known.

The Abbot-Ionescu formulas, as described in Chapter 4, equations (4.61*a–c*), applied to continuity equation (7.1), by considering a weighting factor $\theta = 0.5$, yields:

$$\frac{\partial Q}{\partial x} \approx \frac{1}{2} \cdot \left(\frac{Q_{i+1}^{n} - Q_{i-1}^{n}}{2 \cdot \Delta x} + \frac{Q_{i+1}^{n+1} - Q_{i-1}^{n-1}}{2 \cdot \Delta x} \right) \tag{7.13a}$$

$$b \cdot \frac{\partial h}{\partial t} \approx b_i^{n+\frac{1}{2}} \cdot \left(\frac{h_i^{n+1} - h_i^{n}}{\Delta t} \right) = \frac{\left(b_i^n + b_i^{n+1}\right)}{2} \cdot \left(\frac{h_i^{n+1} - h_i^{n}}{\Delta t} \right) \tag{7.13b}$$

$$q = \frac{1}{2 \cdot \Delta x} \cdot \left(\frac{q_i^n + q_i^{n+1}}{2} \cdot \Delta x + \frac{q_{i+1}^{n} + q_{i+1}^{n+1}}{2} \cdot \Delta x \right) \tag{7.13c}$$

Inserting equations (7.14*a–c*), in (7.1) and grouping all unknown terms, Q^{n+1} and h^{n+1}, at time level $(n + 1)$ on the left hand side of the equation, it yields:

$$A1_i \cdot Q_{i-1}^{n+1} + B1_i \cdot h_i^{n+1} + C1_i \cdot Q_{j+1}^{n+1} = D1_i \tag{7.14}$$

where the coefficients of Q^{n+1} and h^{n+1} are:

$$A1_i = -\frac{\Delta t}{4 \cdot \Delta x} \tag{7.15a}$$

$$B1_i = \frac{1}{2} \cdot \left(B_i^n + B_i^{n+1}\right) \tag{7.15b}$$

$$C1_i = -A1_i \tag{7.15c}$$

$$D1_i = B1_i \cdot h_i^n + A1_i \cdot \left(Q_{i+1}^n - Q_{i-1}^n\right) + \frac{1}{4} \cdot \left(q_i^n + q_i^{n+1} + q_{i+1}^n + q_{i+1}^{n+1}\right) \tag{7.15d}$$

Equation (7.14) is linear in of Q^{n+1} and h^{n+1} and is valid for $i = 1, 3, 5, \ldots, M - 2$. The equation is space-centred around a computational point where h is the unknown value. Because the weighting factor is 0.5 the equation is also centered in time, between the time levels n and $(n + 1)$.

Applying the Abbot-Ionescu formula of discretization for the derivatives of momentum equation, yields:

$$\frac{\partial h}{\partial x} \approx \frac{Q_i^{n+1} - Q_i^n}{\Delta t} \tag{7.16a}$$

$$\frac{\partial h}{\partial x} \approx \frac{1}{2 \cdot \Delta x} \cdot \left(\frac{Q_{i+1}^{n+1} \cdot Q_{i+1}^n}{A_{i+1}^{n+\frac{1}{2}}} - \frac{Q_{i-1}^{n+1} \cdot Q_{i-1}^n}{A_{i-1}^{n+\frac{1}{2}}} \right) \tag{7.16b}$$

$$g \cdot A \cdot \frac{\partial h}{\partial x} \approx g \cdot A_1^{n+\frac{1}{2}} \cdot \frac{1}{2} \cdot \left(\frac{h_{i+1}^{n+1} - h_{i-1}^{n+1}}{2 \cdot \Delta x} + \frac{h_{i+1}^n - n_{i-1}^n}{2 \cdot \Delta x} \right) \tag{7.16c}$$

$$g \cdot A \cdot S_0 \approx g \cdot A_1^{n+\frac{1}{2}} \cdot S_0 \tag{7.16d}$$

$$g \cdot A \cdot \frac{Q \cdot |Q|}{M^2 \cdot R^{\frac{4}{3}} \cdot A^2} = g \cdot A \cdot \frac{Q \cdot |Q|}{K^2} \approx g \cdot A_1^{n+\frac{1}{2}} \cdot \frac{Q_1^{n+1} \cdot Q_i^n}{\left(K^2\right)_i^{n+\frac{1}{2}}} \tag{7.16e}$$

where $K = SM \cdot R^{2/3}$ is the conveyance of the cross-section and SM is the Strickler coefficient.

Equations (7.16*a–e*) are linear, because the term in Q, which was at power two is expressed as a product between a known value at time level n and an unknown value at time level $(n + 1)$.

Similarly, using relations (7.16), for momentum equation, yields:

$$A2_i \cdot h_{i-1}^{n+1} + B2_i \cdot Q_i^{n+1} + C2_i \cdot h_{i+1}^{n+1} = D2_i \tag{7.17}$$

Coefficients of equation (7.17) are:

$$A2_i = -g \cdot A_i^{n+\frac{1}{2}} \cdot \frac{\Delta t}{4 \cdot \Delta x} \tag{7.18a}$$

$$B2_i = 1.0 + A_i^{n+\frac{1}{2}} \cdot \frac{\left|Q_i^n\right|}{\left(K^2\right)_i^{n+\frac{1}{2}}} \cdot \Delta t \tag{7.18b}$$

$$C2_i = -A2_i \tag{7.18c}$$

$$D2_i = Q_i^n - \frac{\Delta t}{2 \cdot \Delta x} \cdot \left(\frac{Q_{i+1}^{n+1} \cdot Q_{i+1}^n}{A_{i+1}^{n+\frac{1}{2}}} - \frac{Q_{i-1}^{n+1} \cdot Q_{i-1}^n}{A_{i-1}^{n+\frac{1}{2}}} \right) \tag{7.18d}$$

Equations (7.14) and (7.17) form together a linear system of equations of the form:

$$\begin{bmatrix} \text{Left BC} & & & & & & & & & & \\ A1_1 & A1_1 & C1_1 & & & \cdots & & & & & \\ & A2_1 & B2_1 & C2_1 & & \cdots & & & & & \\ & & A1_2 & B1_2 & C1_2 & \cdots & & & & & \\ & & & A2_2 & B2_2 & C2_2 \cdots & & & & & \\ . & & & & & \cdots & & & & & \\ . & & & & & \cdots & & & & . & \\ . & & & & & \cdots & & & & . & \\ & & & & & \cdots & A1_{M-1} & B1_{M-1} & C1_{M-1} & . & \\ & & & & & \cdots & & A2_{M-1} & B2_{M-1} & C2_{M-1} & \\ & & & & & \cdots & & & & \text{Right BC} & \end{bmatrix} \begin{bmatrix} Q_1 \\ h_1 \\ Q_2 \\ h_2 \\ \\ . \\ . \\ . \\ \\ Q_M \\ h_M \end{bmatrix} = \begin{bmatrix} \text{Value from BC} \\ D1_1 \\ D2_1 \\ D1_2 \\ D2_2 \\ . \\ . \\ . \\ D1_{M-1} \\ D2_{M-1} \\ \text{Value from BC} \end{bmatrix}$$

The above system of equations can be solved using the classical approach of matrix algebra from calculus or the most used method, the Double Sweep Algorithm, as it was demonstrated previously for Preissmann scheme which uses the physical properties between Q and h to simplify the way the computation is done. As presented in the previous section yhe method of solution is based on continuous replacement from one equation to the next. In equation 1 from the linear system (7.14) the term Q_2 can be transferred in the right hand side in order to obtain an expression for h_1 as a function of Q_2 and other known terms. Further the expression of h_1 is replaced in the second equation and then into the following and so on. Eventually the last equation of the linear system contains one unknown value, the $Q_{N_{x-i}}$. The value of the unknown is calculated and a backward insertion trough the equations can be started in order to determine all the unknowns. Computation is carried on several times, until a stop criterion is fulfilled. Moreover it needs to be noted that the coefficients are expressed using values of Q and h in discretization points where these are not computed, because of the staggered nature of the grid. This issue is overcome by using interpolated values for flow in grid-points where a water depth is computed and vice versa by using interpolated water depth in Q grid-points. The interpolation can be linear or parabolic, and can be calculated as an intermediate value or as a final step in the iteration procedure.

As shown in Chapter 6 the solution accuracy is influenced by the choice of Δx and Δt. Given nowadays computer power the choice of Δx (implicitly of number of discretization points M) is not a problem anymore. Based on several application results as well as theoretical analysis conducted by Wright and Crossato (2013) concluded that the Abbott-Ionescu-Scheme gives numerical stable results without numerical diffusion. Chintu (1986) mention that the Abbott-Ionescu scheme works properly with Courant number values close to 10 due to the implicit nature of the scheme. Compared with an explicit numerical scheme, where the values of Q and A in a point P of the computational domain depends on the values of 3 points from the previous time level, within an implicit scheme as is the Abbott-Ionescu scheme these values depend on 20 points, located to the left and the right of the P point, in the previous time level (see Figure 7.5). The Courant number can be considered as a measure of the extent of the flow regime from time level n to time level $(n + 1)$. The requirement of $Cr < 10$ for the Abbott-Ionescu scheme is equivalent to determine the values for Δx and Δt as:

$$u \cdot \frac{\Delta t}{\Delta x} < 1 \tag{7.19}$$

$$\frac{\Delta h}{\Delta x} \cdot \left| u \pm \sqrt{g \cdot h} \right| < 10 \tag{7.20}$$

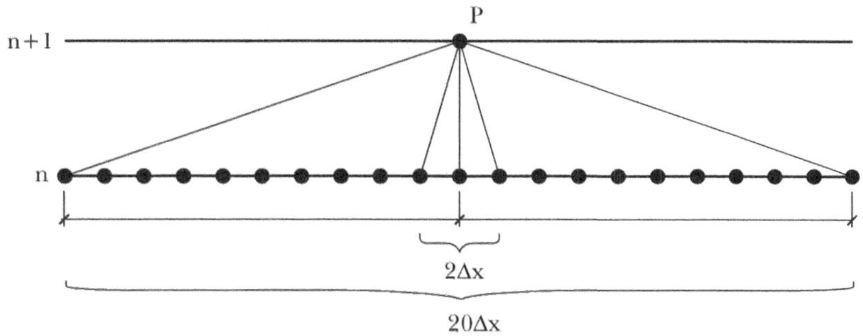

Figure 7.5 Domain of dependence for a point P in the computational domain, when Abbott-Ionescu scheme is used for solution of Saint-Venant equations.

The values of Δx and Δt as defined by equations (7.19–7.20) have to be correlated with the variations in initial and/or boundary conditions and of the tributaries of a modelled river. The choice of Δx and Δt should make a good description of all variations in times series as well as longitudinal profiles.

7.2.3 Initial and boundary conditions

Modelling river channels give good results if appropriate initial and boundary conditions are set to the model. Initial conditions are easy to set, and they usually come with the knowledge about the modelled system. These are given in terms of water levels or discharge values.

Boundary conditions that are usually set at the upstream end of a river are water level or inflow variations in time. A rating curve should never be specified as a boundary condition at the upstream end of a model. Boundary conditions at the downstream end of a model are water levels, outflow hydrographs and rating curves. Cunge *et al.* (1980) advice that the same type of boundary conditions (Q or h) at both river ends is not recommended, because solution strongly depends of the initial conditions. The most used boundary conditions at the downstream end should be measured Q and h as variations in time. If discharge and/or water levels variations in time are not available at a downstream end of a river reach a rating curve ($Q - h$ relationship) is used. However, such an approach may give problems, because it is represented as a single valued function. In practice a rating curve has hysteresis for unsteady flow conditions. In order to eliminate the influence of a rating curve Singh *et al.* (1996) suggests that models should be artificially extended in the downstream.

7.2.4 River networks

It is important to mention that in case lateral inflow is considered in Saint-Venant equations, then relation (7.14*c*) takes into account the possible variations in space and time by considering the values of q from space point i to space point $(i + 1)$ and values at time levels n and $(n + 1)$. Notice that lateral inflow variation in time and space has to be known when applying Preissmann or Abbott-Ionescu schemes. If the model has a network of river branches that are joining together, than a tributary is a lateral flow to the main branch. Figure 7.6 gives the possible variation of q in space at each time step. The flow values coming from a tributary at time level n ($t^n = n \cdot \Delta t$) are added to the lateral inflow term. In case that Abbot-Ionescu scheme is used for the solution of the Saint-Venant equations due to the staggered nature of the discretization, grid tributaries have to be associated with odd i-points and they should be dispersed over a distance $2 \cdot \Delta x$. This approach requires careful selection of Δx.

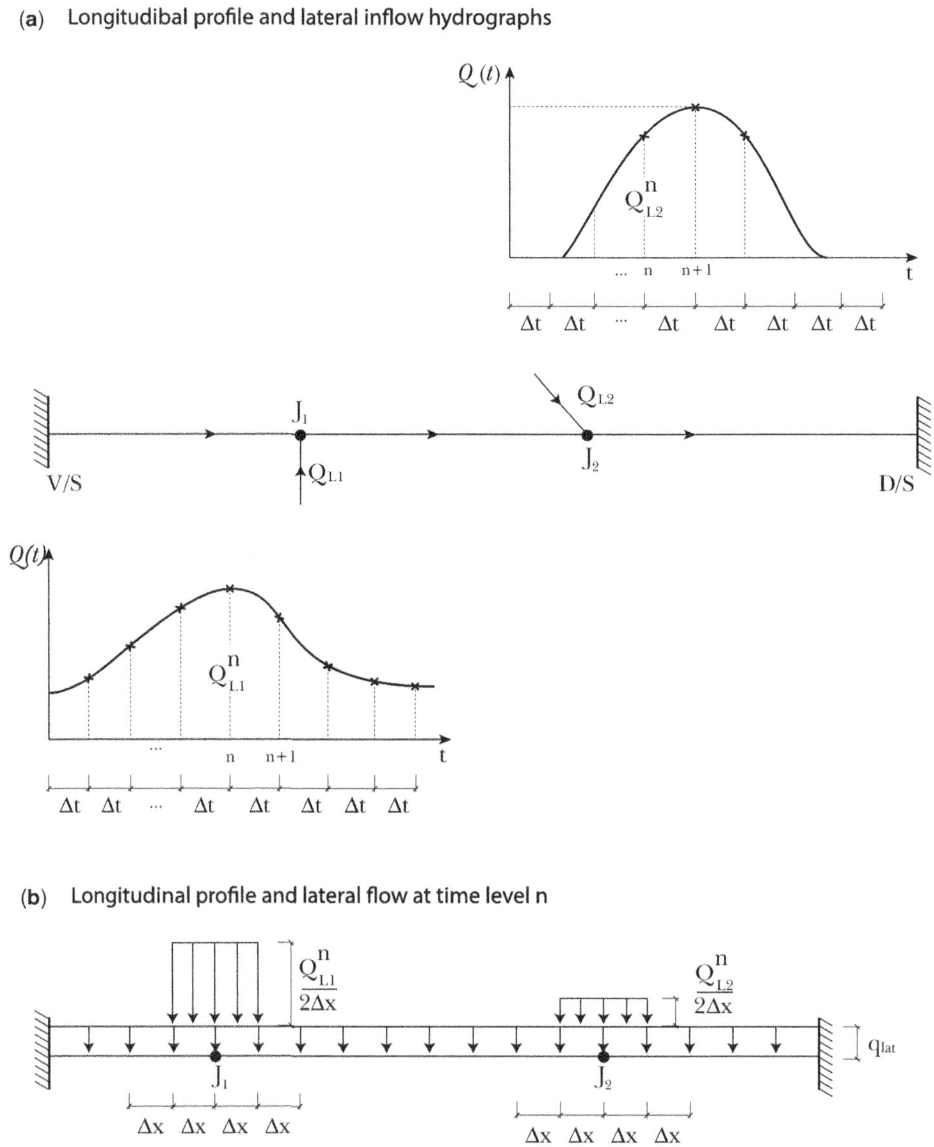

Figure 7.6 (a) Lateral inflow variation for network river systems and (b) tributaries handling in the Abbot-Ionescu solution approach.

In case a river is formed from a network of channels the same procedure of solution is applied as for a single river reach, provided that boundary conditions are specified at all ends of the pendant branches that is in points 1, 71 and 101, as per Figure 7.1. A system of algebraic equations that need to be solved is obtained after adding additional equations for the channel junctions (see Figure 7.6):

$$\begin{cases} Q_i = Q_j + Q_k \\ h_i = h_j \\ h_i = h_k \end{cases} \tag{7.21}$$

First equation in conditions (7.22) represents continuity equation, whereas the later two are simplified forms of energy equation written at the junction, under the assumption that local losses are neglected. Forsius and Huttula (1982) show that if flow velocities are too high, such that velocity head becomes significant, the equal water level condition in (7.22) should be replaced with an equal energy level compatibility equation.

The corresponding matrix of the solution system for the Saint-Venant equations is represented in Figure 7.8, for the river network of Figure 7.7.

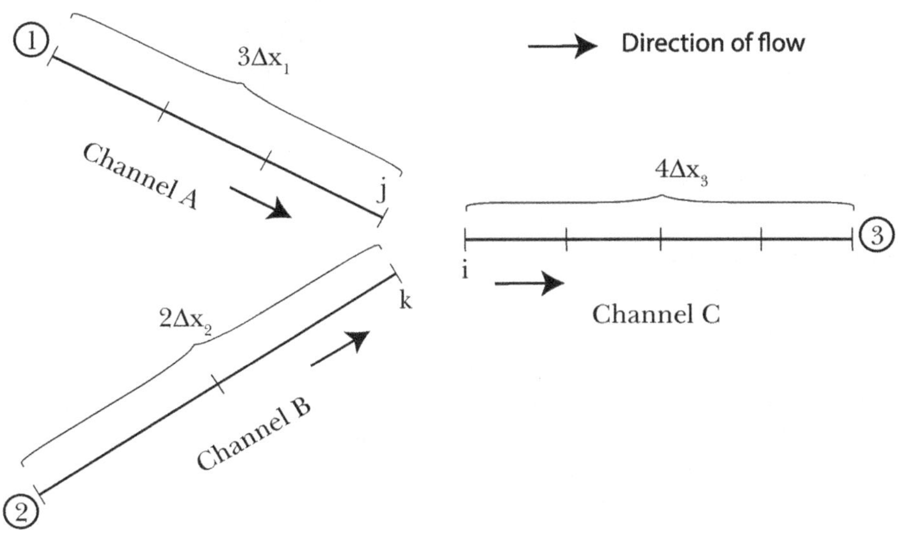

Figure 7.7 Channel junctions.

The double sweep method can also be applied to network river systems as long as certain computational order is respected (Wu, 2008). For the channel network of Figure 7.7, the computational order is as follows; the forward sweep computes the recurrence coefficients for channel A starting from 1, where boundary condition is known, and is carried on down to junction point j. A similar sweep is carried on channel B from first to last point. At the junction the three cross sections should be located very close to each other and according to relation (7.21) it is assumed that the water levels at the cross sections are equal and discharge at the downstream cross section is equal to the sum of discharges in upstream cross sections of channel A and channel B, hence initial values for coefficients F and G for channel 3 are computed. The forward sweep is further carried out from point *i* to point 3 of channel C and the recurrence coefficients are computed. Backward sweep start from the most downstream point of channel C (point 3), where boundary condition are known. Values for discharge and water level are computed first along channel C back to the junction. From the junction the backward sweep algorithm is carried on further along channel A and B.

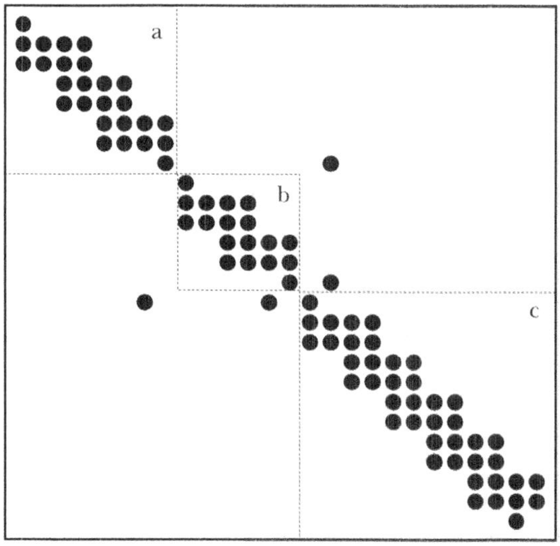

Figure 7.8 Structure of a matrix for the river network system represented in Figure 7.7.

7.3 MODELLING FLOODS

Flood are phenomena characterised by high discharges and/or water levels, which may cause inundation of floodplains (the land adjacent to rivers) or to terrain close to water bodies. In general flooding happen when rivers are unable to route the amount of water coming to the river due to runoff. Because a river has no conveyance capacity leads to overtopping of river banks. Floods causes disruption of activities and loss of lives, hence they are given a lot of attention in water engineering practice. Flood disasters account for about a third of all natural disasters (by number and economic losses).

Flood events may be caused by intense or long-lasting rainfall, snowmelt, dam or dyke break, ice jams, high tides, storm surges or by operation of reservoirs.

One dimensional models for Saint-Venant applications have been widely applied in simulating flood routing and unsteady flow in river networks (Hicks *et al.*, 2005; Gichamo *et al.*, 2012; Moya Gomez *et al.*, 2013). Though there are advances in the two (2D) and three dimensional (3D) hydrodynamic modeling of flow in rivers, models based on one dimensional Saint-Venant equations still are the most popular choice for solving large-scale river engineering problems. The choice is due to their reduced data requirements compared to those for 2D and 3D models, as well as reduced computational cost.

Special insights in flood propagation in river channels are gained by using Price's (1985) analysis of the Saint-Venant equations (7.1), which can be rearranged in the form:

$$\frac{\partial Q}{\partial t} + \frac{Q}{b}\left[\frac{1}{K}\frac{dK}{dh} - \frac{1}{2b}\frac{db}{dh}\right]\left(\frac{\partial Q}{\partial x} - q\right) - \frac{K^2}{2bQ}\frac{\partial^2 Q}{\partial x^2} = 0 \tag{7.23}$$

Equation (7.23) is obtained by neglecting the inertial terms (the local acceleration and convection of momentum), and has the form of an advection-diffusion equation with advection and diffusion terms:

$$c(Q,h) = \frac{Q}{b}\left[\frac{1}{k}\frac{dK}{dK} - \frac{1}{2b}\frac{db}{dh}\right] \tag{7.24a}$$

and

$$D(Q,h) = \frac{K^2}{2bQ} \tag{7.24b}$$

Coefficients given by relations (7.24) show that when a flow disturbance occurs on a river channel, a process of translation/advection takes place along with a corresponding diffusion. The phenomenon is known as attenuation of peak flow. The speed of the wave traveling downstream is given by the advection speed (7.24*a*). Note that equation (7.23) has Q as the primary dependent variable within the differentials, however because Q is dependent on h (or A), a continuity equation is required in order to solve it.

An more general representation of equation (7.23) given by Price (1985) is:

$$\frac{\partial Q}{\partial t} + c_0\frac{\partial Q}{\partial x} + \varepsilon c_0\frac{\partial}{\partial t}\left(\frac{a_0}{c_0^2}\frac{\partial Q}{\partial x}\right) = c_0 q - \varepsilon\gamma c_0\frac{\partial}{\partial t}\left(\frac{Qq}{2qAs_0}\right) + O(\varepsilon^2) \tag{7.25}$$

where

$$\varepsilon = \frac{\frac{\bar{h}}{\bar{x}}}{s_0} \tag{7.26a}$$

$$\gamma = \frac{\bar{Q}^2}{g\bar{A}^2\bar{y}} \tag{7.26b}$$

$$a_0 = \frac{Q}{2s_0 B}\left[1 - \frac{\gamma B}{gA}\left(c_0 - \frac{Q}{A}\right)^2\right] \tag{7.26c}$$

Equation (7.25) is the basic flood routing equation, valid for any Froude number. It is also known as the Muskingum-Cunge equation. Muskingum is the name of a river in North America and Cunge is the scientist who first established the relationship between the Muskingum method and the Saint-Venant equations (Cunge, 1969). Based on equation (7.25) attenuation parameter can be determined for any river for specific discharge values. An example of a typical such curve is given in Figure 7.9.

Flood attenuation studies (Anderson *et al.*, 2006; Acreman *et al.*, 2003; Castro Gama *et al.*, 2014) concluded that the peak discharge attenuation is affected by floodplain (width and roughness) as well as by bed slope. Hence floodplain storage is an important process to take into consideration while predicting the peak flood attenuation. One-dimensional models should be capable to account for floodplain storage during flood events, hence, considering a coupled 1D/2D model in flood simulations is important. An extensive overview of flood inundation models is given by Horrit and Bates (2003).

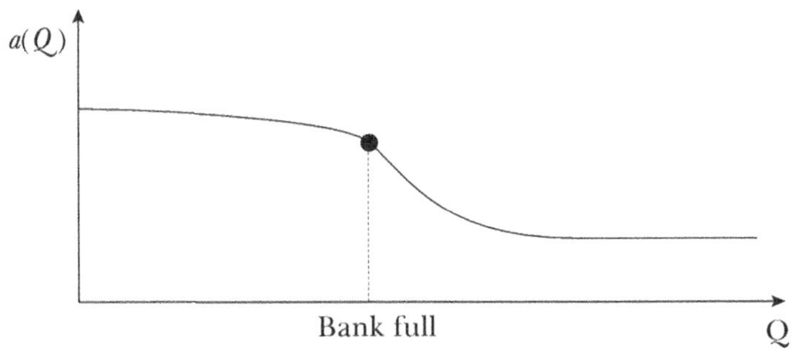

Figure 7.9 Typical flood attenuation curves.

7.4 RIVER ROUTING EXAMPLE

In order to show the capability of a river to route flow the example of a simple rectangular canal is chosen. A flood hydrograph is routed downstream on the 200 km long channel in different conditions. Different widths of the rectangular cross-section are considered, as well as different slopes and roughness. HEC-RAS is used (i.e., Preissmann scheme) to compute the different cases. The hydrograph that is routed downstream the considered canal is shown in Figure 7.10, along with the schematic of the canal geometry.

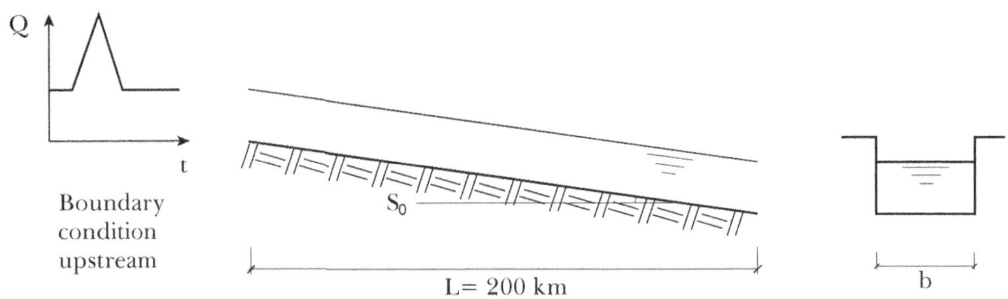

Figure 7.10 Canal geometry and boundary conditions.

Results of routing are shown in Figures 7.11, 7.12, 7.13. Figure 7.11 shows the effect of roughness in flood routing. Figure 7.12 shows the effect of the slope on river floods. Steep slope route quicker the flood peak, however the attenuation is smaller. Figure 7.13 shows the effect of cross-section width on river routing. The wider the river cross-section the higher the attenuation of the flood peak, however longer the time for the flood to be routed out of the river system.

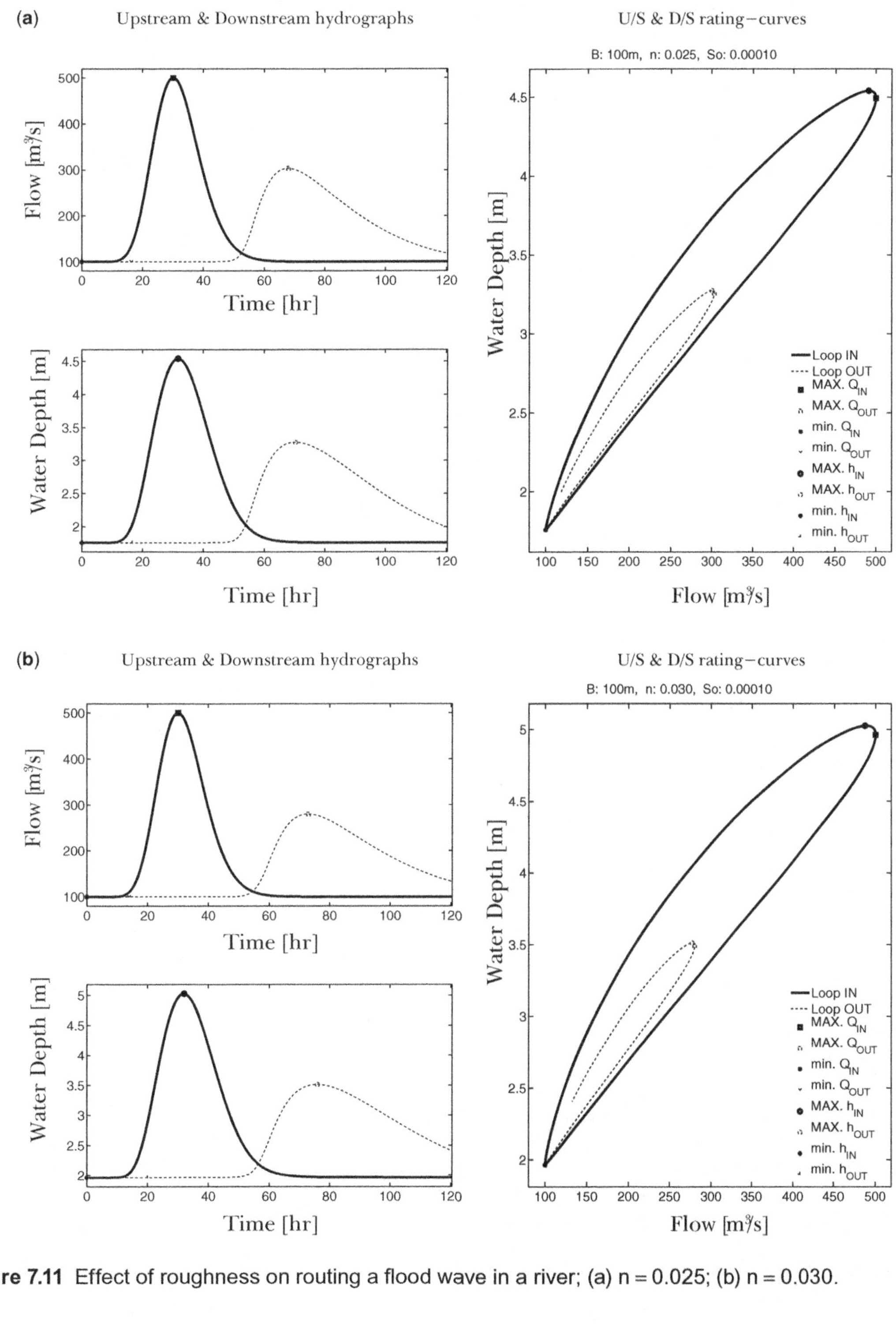

Figure 7.11 Effect of roughness on routing a flood wave in a river; (a) n = 0.025; (b) n = 0.030.

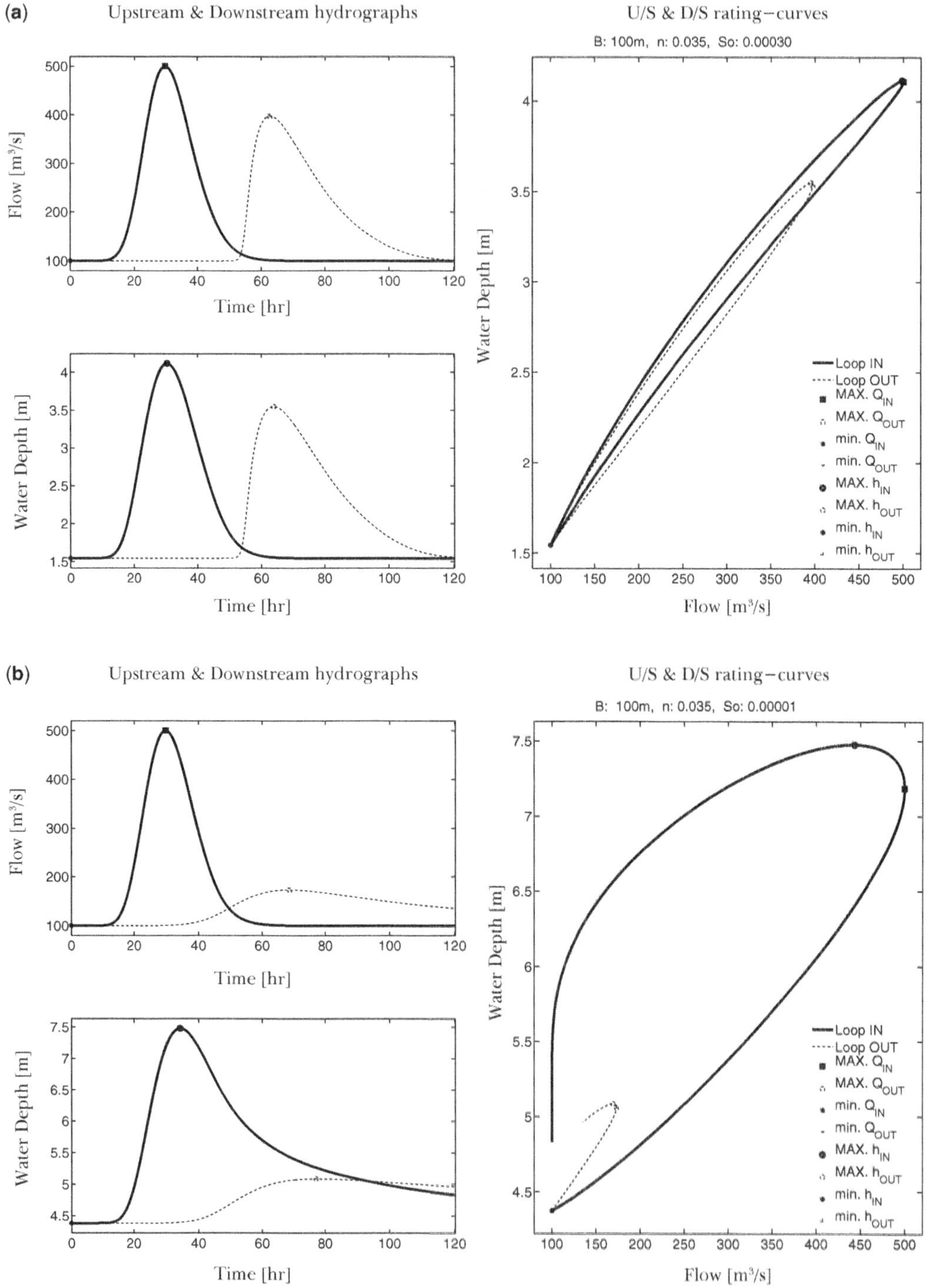

Figure 7.12 Effect of slope on routing a flood wave in a river; (a) S = 0.0003; (b) S = 0.00001.

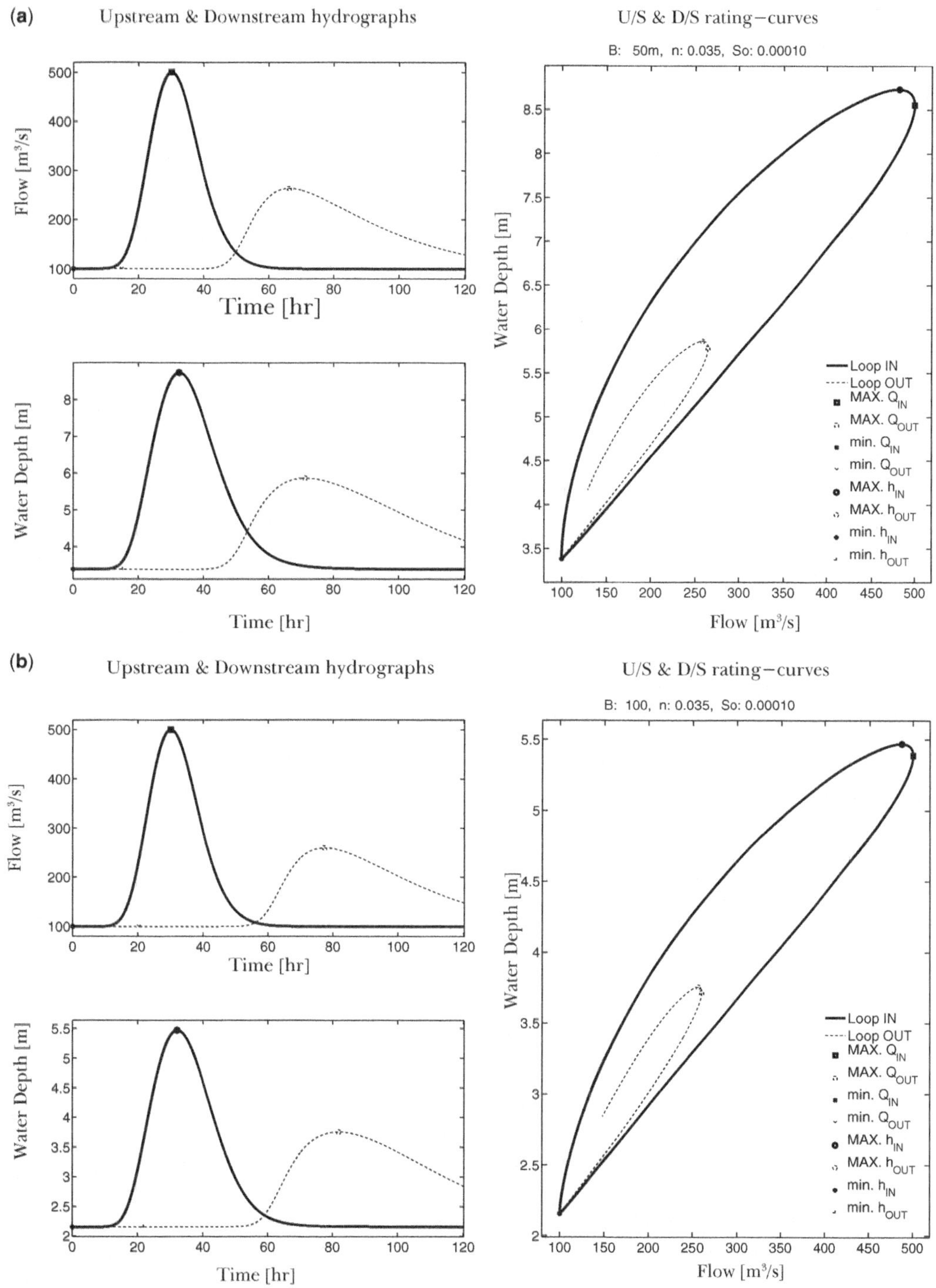

Figure 7.13 Effect of cross-section width on routing a flood wave in a river; (a) 50 m width; (b) 100 m width.

Chapter 8

Water quality modelling

8.1 INTRODUCTION

Fresh water is very important for development because it is a source of energy, a mean for transportation and a habitat for aquatic life. Availability of good quality water (fresh water) is very important for economic activities and growth, however as population and economies grow, so does the amount of pollutants that are discarded in water bodies (Orlob, 1983). When levels for water quality are not met, treatment must be carried out, which brings extra costs to water users.

The issue of water quality is not something new, as Heathcote (2009) notes Romans issued the following Roman decree:

> *'Ne quis aquam oletato dolo malo ubi publice saliet si quis oletarit sestertiorum X mila multa esto.'*

which reads

> 'It is forbidden to pollute the public water supply: any deliberate offender shall be punished by a fine of 10,000 sesterces'. Year 1989 was the 2000th anniversary of the issuing of this decree.

Countries around the world defined standards for water quality and make plans for reaching good ecological status of their water resources. A good example of such efforts is the European Water Framework Directive, which is a major policy in the area of water resources management in European Union (EU). All countries of EU follows the policy, which sets clear requirements regarding attitudes and approaches to water resources at both national and regional levels. The directive's main goal is sustainable use of water resources with a dominant 'environmental objective' of preventing the deterioration of the status of all EU waters.

In this context of continuous need to maintain and improve water quality scientists and managers are continuously trying to come with solutions for providing understanding of freshwater ecological systems and predictions of the water quality responses to human interventions, as well as evolution of these responses in time. Hence water quality models are important tools assisting in the analysis regarding pollution of water and ultimately in the identification of the state of the environment. Such tools need to be able to show the impact of humans activities on the status of a water body (i.e., lake, river, etc.).

Water quality modelling aims to simulate behaviour of variables related to quality of a water body. Variables include characteristics of the water body, such as temperature, solar radiation, wind speed, as well as pollutants existent in the modelled water body. Models of water quality describe physical processes in terms of hydrodynamics; and are complemented by a description of chemical and biological processes that control the transport and transformation of the variables. Pollution elements are defined as point-source and nonpoint source loads. Developed water quality models range from very general, applicable to all kind of water bodies; to very specific ones, that are applicable to a particular problem type. Hence each model may have its own characteristics and requirement for data input. Water quality models are known not to be able to accurately predict what happens, however they are still valuable to understand processes and estimate relative changes due to management practices and policies. Hence even the simpler water quality model is a useful tool to be used.

First water quality model, the so-called S-P model, was defined by Streeter and Phelps (1925) for Ohio river, in United States. The S-P model is steady state one-dimensional determining the oxygen balance and the decay of bio-chemical oxygen demand (BOD). Since 1925 models have evolved from steady state models to dynamic ones and from point source pollution to non-point pollution sources (Jung *et al.,* 2011), as it is the example of EFDC model developed by Virginia Institute of Marine Science, which is suitable to be used for water quality simulation in rivers, lakes, reservoirs, estuaries, and wetlands, for one-, two-, or three-dimensions (EPA, 1999). Table 8.1 presents the main available models and their characteristics, as they evolved in time.

Table 8.1 Water Quality Models.

Model	Year of development	Main characteristics
Streeter-Phelps	1925	– One-dimensional – Steady state – Oxigen balance
QUAL	1970	– One-dimensional or dynamic – Non-point source pollution – Suitable for dendritic rivers
WASP	1983	– One-, two-, or three-dimensional models
DHI models	1990	– One-, two-, or three-dimensional models; – Non-point source pollution; – Suitable for rivers, lakes, estuaries, and reservoirs;
BASINS	1993	– One-, two-, or three-dimensional models; – Non-point source pollution; – Suitable for catchments;
EFDC	1997	– One-, two-, or three-dimensional models; – Non-point source pollution; – Suitable for rivers, lakes, estuaries, and reservoirs

An important aspect of water quality modelling is the availability of monitoring data, which decide the appropriate model to be used for the complexity of the situation (Louks & Beek, 2005). Depending on the availability of monitored water quality data it is advisable to begin with simple models and gradually extend to additional complexity as data becomes available due to data collection campaigns. Prediction of hydrodynamics of the water body has long been the recognised as an important component of any water quality study (Martin & McCutcheon, 1999). It is well known that river have good collection data for hydrodynamic parameters, however most water quality monitoring programmes for water bodies with stagnant waters, detailed hydrodynamic measurement is hardly included neither is prediction of water motion, the emphasis is mainly on general physical-chemical parameters due to their ease of measurement. In case of lakes water quality parameters collected during spring and fall overturn represent the best and most uniform water quality conditions and are most valuable for calibrating and validating models. However making model predictions based on one or two samples is not enough. More extensive sampling is suggested because it provides additional information.

Present chapter presents the main equations that are used to define water quality models. First the main processes described by a water quality model are presented, followed by specific particularities of processes for rivers and lakes. Models used for lakes and reservoirs are substantially different from the ones used for rivers. Usually water movements are very slow in lakes and reservoirs, hence the hydrodynamics is driven by wind; heat exchange due to solar radiation; and inflows and outflows. Phenomena that are normally absent or negligible in river flow are very important in non-flowing waters ('standing' waters) such as lakes and wetlands. For example in lakes the phenomena of temperature stratification appears, which can influence the vertical turbulent mixing; or different density stratification of water causes advection. These phenomena are known to create a particular thermo-hydrodynamics state in a lake. In rivers the main driving force is gravity and bed friction (as resisting force), which generates perceptible velocities and turbulent mixing. Moreover sediment transport and deposition is different in rivers and lakes, therefore their effect on water quality is different. Usually in lakes and reservoirs a phenomenon of slow deposition of fine sediments takes place and the bottom of the lake can release substances, as effect of bacterial decompositions and chemical processes that are taking place. In rivers such processes are negligible as compared with the advective transport of substances in the water.

8.2 PROCESSES DESCRIBED IN WATER QUALITY MODELS

Water quality models are applied to different types of water bodies; rivers, estuaries, lakes, reservoirs, coastal waters and seas or oceans. Models are usually used to describe the main water quality processes, and require input in form of hydrological time series of inflow and outflows, as well as pollutant loading (pollutant concentrations).

Water quality models are a set of equations that describe the advective and dispersive transport of pollutant constituents within the water body, based on the hydrodynamic conditions; as well as a set of equations describing physical, chemical and biological reactions among constituents.

Advective transport is dominant in river, where flow takes place, while dispersion is predominant transport phenomenon in estuaries, which are under tidal influence. Lakes and reservoirs water quality is more difficult to predict than the one in rivers because of the wind direction, which is random in nature. Currents formed due to wind influence affect surface mixing and generate stratification of the water.

The basic principle of water quality models is application of mass balance. Space computational domain, is formed by one-, two-, or three-dimensional system of the water body (i.e., river, lake, estuary, a zone in the sea or the ocean). Space and time domain is divided, using a discretization grid, in computational cells (volume elements) in space and in discrete time intervals Δt. Water quality is computed over time for a computational cell volume by writing the mass balance for each water quality constituent. Flow characteristics in the control volume, over the time interval Δt, is pre-computed as a hydrodynamic solution carried out for the given inflow and outflow conditions.

For each computational cell and each time step Δt, the mass balance of a substance can be defined as a sequence of step by step computations of:

- changes due to transport in and out of the computational cell (TR), which can have two components; and advective term (flowing water) and a dispersive term (pollutant concentration differences);

- changes due to physical or (bio)chemical processes (P) occurring within the computational cell; and
- changes attributed to sources/discharges to or from the computational cell (S).

Mass balance over a time step interval Δt, for computational cell i at time level $t + \Delta t$ can be expressed as:

$$M(t + \Delta t) = M(t) + \int_{t}^{t+\Delta t} M_{TR} dt + \int_{t}^{t+\Delta t} M_P dt + \int_{t}^{t+\Delta t} M_S dt \tag{8.1}$$

In equation (8.1) M is the mass in computational cell i, M_F is the mass change in computational cell i, due to component F, which is transport (TR); (bio)chemical or biological processes (P); and sources (S).

Equation (8.1) of the mass balance for computational cell i at time level $(n + 1)$ can be further expressed as (Louks & van Beek, 1983):

$$M_i^{n+1} = M_i^n + \Delta t \cdot \left(\frac{\Delta M_i}{\Delta t}\right)_{TR} + \Delta t \cdot \left(\frac{\Delta M_i}{\Delta t}\right)_P + \Delta t \cdot \left(\frac{\Delta M_i}{\Delta t}\right)_S \tag{8.2}$$

where M_i^n is the mass in computational cell i at time level n and M_i^{n+1} is correspondingly the mass in computational cell i at time level $(n + 1)$. The term $\left(\frac{\Delta M_i}{\Delta_t}\right)$ represent the changes in computational cell i, due to component F.

Transport due to dispersion is important in lakes where water column is stratified, and it is less visible in rivers. Dispersion is not the same as diffusion. Figure 8.1 shows schematically possible transport elements in lakes and rivers.

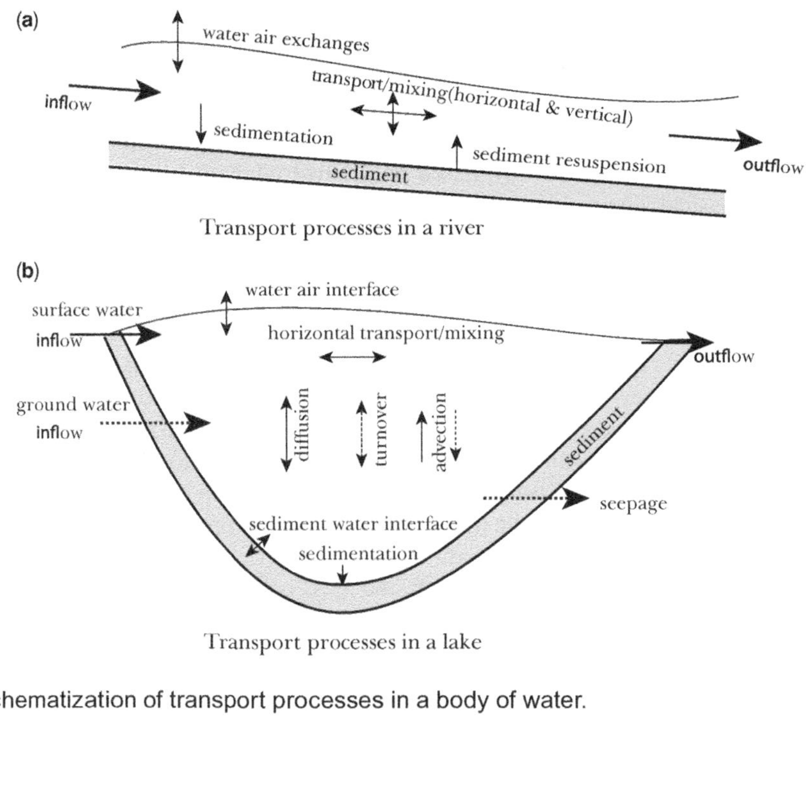

Figure 8.1 Schematization of transport processes in a body of water.

Physical processes include settling of sediments and substances, as well as phenomena such as aeration. Biochemical processes transform one substance into another through adsorption, decay, or biological processes. Sources add or subtract mass from the computational cell. Additional pollutant mass can come due to sewer wasteloads and subtraction is due to intakes.

The three change processes described above (TR, P and S) are governed by the extended transport equations (Somlyódy & van Straten, 1986), which is written in three dimensions as:

$$\frac{\partial C}{\partial t} = \frac{\partial}{\partial x}\left(D_x \cdot \frac{\partial C}{\partial x}\right) - \frac{\partial}{\partial x}(u_x C) + \frac{\partial}{\partial x}\left(D_y \cdot \frac{\partial C}{\partial x}\right) - \frac{\partial}{\partial x}(u_y C) + \frac{\partial}{\partial x}\left(D_z \cdot \frac{\partial C}{\partial x}\right) - \frac{\partial}{\partial x}(u_z C) + rc(\boldsymbol{C}, param) \quad (8.3)$$

where $\boldsymbol{C}(c_1, c_2, \ldots, c_n)$ is the n dimensional mass concentration vector for n constituents of analysed water quality; $\boldsymbol{u}(u_x, u_y, u_z)$ is the velocity vector; $\boldsymbol{D}(Dx, Dy, Dz)$ is a vector containing diffusion coefficients for the x, y and z, directions respectively; and $rc(\boldsymbol{C}, param)$ is the n dimensional vector of rates of change of constituents due to P and S processes as a function of concentrations of the n constituents and a number of parameters, *param* that depend on the model choice. Model calibration is usually performed to determine the values of *param*.

In case sediment is a component of the water quality model a mass conservation equation for the sediment is added to the model along with corresponding interface terms sediment-water and water-air.

Equation (8.3) is a system of PDEs that are solved numerically (together with the sediment equation in case sediment is considered). Another approach to solve the formed system of PDEs is to use a number of m interconnected conceptual elements (so-called tanks) that lead to a system of $n \times m$ ODEs.

Models that are defined in steady state are the ones for which there are no changes in the concentrations with time, that is, the term $\partial C/\partial t$ in equation (8.3) equals zero.

Temperature affects water quality processes, hence it is important to incorporate temperature and its modelling when it varies over the simulation period, or when the inflow to the water body has a heat that is different from the one of the water body that is modeled. Temperature models are based on heat balance of the modeled water body, by taking into account solar and atmospheric radiation and evaporation. Such models are not easy to set-up, because they require a lot of measured data that is not always available.

8.3 RIVER WATER QUALITY MODELS

River waters are usually receiving pollutants from several sources such as nutrients carried by sediments discharged in a river by runoff; sewers and storm sewer overflows; and so on. Mathematical models are useful in describing the state of water quality in a river system, as well as to predict the change in state, in case initial and/or boundary conditions are changed. Examples of changes are: changes in the type (point or non-point) of loads; morphological changes of the river bottom, due to construction and/or operation of control structures (weirs, dams, etc.); or changes in hydrograph inputs (upstream and lateral) due to effect of climate change. The degree of complexity in describing river water quality state varies from model to model. Initially, river water quality models determined dissolved oxygen concentrations (DO) and, later were extended to predict plankton-nutrient dynamics in river eutrophication problems (Rauch *et al.*, 1998). Other constituents that may be incorporated in water quality models are macrophytes (Barendregt & Bio, 2003; Zalewski, 2004); phosphorus; oil spills and pesticides. A comprehensive overview of existing river water quality models is given by Cox (2003).

Oxygen concentration of river waters is the most important factors affecting aquatic life. Hence water quality standards are related to the concentration of dissolved oxygen (DO) and parameters affecting it. Discharging pollutants in receiving river waters has as effect a reduction in the oxygen content. A related element is the Biochemical Oxygen Demand (BOD), which shows the oxygen consumption due to degradation of organic material. BOD5 is the amount of BOD consumed over 5 days.

Oxygen consumption is described in form of a first order ordinary differential equation:

$$\frac{dC_{\mathrm{BOD}}}{dt} = -(K_1 + K_{1,S}) \cdot C_{\mathrm{BOD}} + K(C_{os} - C_o) - D_b \tag{8.4}$$

where CBOD is the concentration of BOD (M/L3); K1 the decay constant (sec^{-1}); K1S the sedimentation/adsorption constant (sec^{-1}); Cos the oxygen saturation constant (M/L^3); C_o the oxygen concentration in the river (M/L^3); K the re-aeration kinetic constant (sec^{-1}) and D_b the rate of oxygen removal caused by benthic processes. Consequently oxygen concentration in a river is determined using the equation:

$$\frac{dC_O}{dt} = -K_1 \cdot C_{\mathrm{BOD}} \tag{8.5}$$

An important phenomena in modelling water quality in rivers is re-aeration that occurs at the air-water interface:

$$\frac{dR_0}{dt} = K(C_{os} - C_o) \tag{8.6}$$

Removal of organic material due to sedimentation process (Dobbins, 1964) is represented in equation (8.4) by the sedimentation/adsorption constant. The actual amount of BOD sedimented/adsorbed at the bottom of a river was described by Harremoes in 1982, which is an important component of river water quality analysis when storm sewer overflows are involved because of their high content in suspended organic content ready to be settled.

Oxygen concentrations are important for the aquatic life, however other factors are also important, such as is the case of ammonia in non-ionized form (NH^3), which is toxic to fish. The pH value in a river determines the dissociation of ammonia into ionized and non-ionized, the higher the PH the higher fraction of ammonia is non-ionized. At pH values above 9 almost all ammonia is in a non-ionized form. Values of pH should not be too high, neither too low, because high pH shows increasing acidification, while low pH is toxic to the aquatic life, because alluminium's solubility increases.

8.4 LAKES WATER QUALITY MODELLING

The above presented water quality modelling concepts are applicable as well to lakes, coastal or ocean waters. This section presents just few aspects that are specific to water quality modelling in lakes and reservoirs. Usually lakes and reservoirs are studied as water bodies with the same physical properties, despite their minor differences. Reservoirs are a result of a built structure (most often a dam), and are used primarily to address problems of water shortages, excess floods, or energy production (ILEC, 2005).

If a river flows into a lake there are significant changes in the state of its water quality because its velocity reduces. All materials (pollutants and sediments) are carried at higher velocities in rivers and they enter to slow moving waters. In lakes and or reservoirs there are more opportunities for algae (phytoplankton) growth and development of eutrophication.

Water quality in lakes and reservoirs is assessed based on several characteristics of the water body. These were defined by Louks and van Beek (1983) as:

- clarity or transparency of the water, that is, greater water clarity indicates better water quality;
- concentration of nutrients, that is, lower concentrations show better water quality;
- quantity of algae, that is, lower levels of algae indicate better water quality;
- oxygen concentration, that is, the higher the concentration of oxygen the better for fish;
- concentration of dissolved minerals, that is, low values give better water quality;
- pH value (a neutral pH of 7 is advised).

Supply of water to a lake is important for determining its water quality. There are two main sources of water supply for lakes and reservoirs; precipitation and river inflow. If precipitation is the major source of water, the lake is acidic and low in nutrients. If river inflows are the major source for water supply, lake water quality is more variable, and has high nutrients level. For the later case water quality depends on the volume of inflow and runoff from the surrounding slopes. Human activities around a lake have important effect on its water quality.

Lakes and reservoirs have three fundamental characteristics that make them quite different in physical behaviour as compared with rivers. These characteristics include: mixing, long retention time, and response dynamics.

Reservoirs or lakes receive inputs from diverse sources; precipitation, river inflow, heat and wind-induced energies that cause waves, thermal energies, contaminants, nutrients and organic substances; which are mixed within a lake so that both the resources and problems are dissipated throughout its volume.

Lakes and reservoirs are generally slow to respond to changes due to their long retention times. They store floods, pollutants, and heat without experiencing immediate changes. The implications of the long retention times is that lakes are relatively stable; allow for suspended materials to settle to the bottom, acting as high efficiency sinks for many materials; and allow for complex, and sometimes unique ecosystems to evolve. The drawback of long retention time is that, once a lake or reservoir is degraded, it takes a long time for it to recover or be restored.

The response dynamics of lakes to a nutrient concentration is different from one of a river. Lakes and reservoirs respond to changes in a highly non-linear way, as depicted in Figure 8.2. Degradation in response to increased nutrient concentrations (from A to B), may not be visible until nutrient concentrations are high and the water body abruptly switches its status. In Figure 8.2 the plankton concentrations are high, and the lake goes from status B to C. The lake cannot simply go back from C to B. There are likely to be irreversible changes to the ecosystem, so recovery follows a path from C down to A through D.

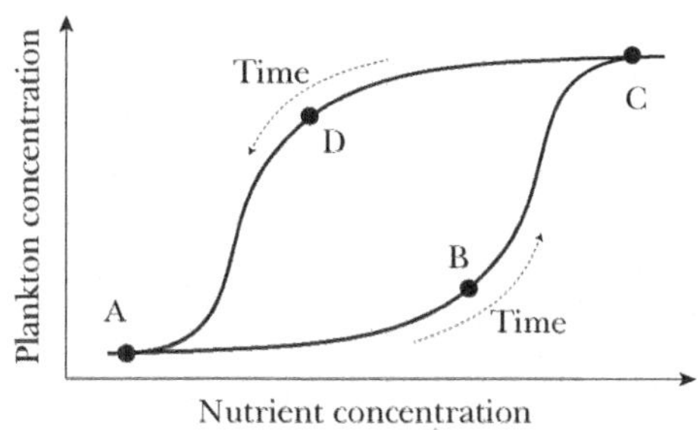

Figure 8.2 Response of a lake to eutrophication process.

Water quality of a lake is affected by the extent to which the water mixes. The depth, size and shape of a lake are the most important factors influencing mixing, along with other factors such as climate, lakeshore topography and inflow from rivers. Density of water has highest value at 4°C. Variations in density caused by difference in temperatures prevent warm and cold water from mixing. Uniform water density facilitates complete mix of lake waters, recharging the bottom water with oxygen and bringing nutrients to the surface. Wind and waves circulates warm water only 6–8 m deep. If a lake is shallow the water may stay completely mixed for the entire summer season. During summer season deep lakes experience stratification into three zones: epilimnion (warm surface layer), thermocline or metalimnion (transition zone between warm and cold water), and hypolimnion (cold bottom water). Stratification traps nutrients from bottom sediments in the hypolimnion. In the fall season, the lake surface cools and eventually water temperature becomes uniform from top to bottom, hence mixing appears (fall overturn). During fall season often algae bloom appears to the surface of a lake because of the mixing.

8.5 EXAMPLES OF LAKE HYDRODYNAMICS AND WATER QUALITY MODELS

8.5.1 Sontea-Fortuna wetland system

The Danube Delta is the largest wetland of the Danube river system (see Figure 8.3). It consists of a network of river channels, lakes, and wetlands that help to improve water quality of the receiving waters of the Danube, the Black Sea. Danube Delta forms a buffer zone that filters pollutants from the Danube river, while helping to reduce flood peaks by storing water. A management plan of the Danube Delta is in place in order to ensure its protection and to establish the preservation of an ecological equilibrium. Wetlands of the Danube Delta are part of this management plan, however their functioning is difficult to be quantified because of the dynamic nature of the system (Popescu *et al.*, 2014). Status of functioning of a wetland in the Danube Delta is evaluated based on a clear defined number of indicators (EU WFD, 2000). These indicators are used to assess the condition of the ecosystem, by monitoring trends in time and space. Sontea-Fortuna wetland is one of the most important wetlands of the Danube Delta system because of its location at the entry point in the delta (see Figure 8.4). Terrestrial and aquatic habitat of this wetland depends on the water level in the Sontea-Fortuna area. The major driving force for changes in the system is the hydrological regime of the canal system of the wetland, including periodic floods.

The Sontea-Fortuna wetland is located on Romanian part of Danube Delta and covers a surface of 24,636 ha between Chilia branch to the West and North; Sireasa and Stipoc canals to the North; Catavaia canal and Old Danube to East; and Tulcea and Sulina branches to the South (see Figure 8.3). The area elevation of the wetland varies from as low as –6 m m.b.s.l to +4.5 m m.b.s.l. The lowest elevation in the area occurs along Sontea canal as the central axis of the Sontea-Fortuna system.

Results of a 2D structured grid hydrodynamic model are shown here. The model is used is to determine the behaviour of the system during high and low flow conditions, in order to predict the hydrological regime in the Sontea-Fortuna wetland. The aim of the study is to determine the area exposed to dry conditions, especially during low flows. Three indicators shows the status of the Sontea Fortuna wetland; the inundation pattern, which is in a good status if the water level in the wetland ranges between 1.5 m m.b.s.l and 3.5 m m.b.s.l; flooding duration, that should not exceed 30 days and circulation of water in Fortuna lake, which should not be in a stagnant state.

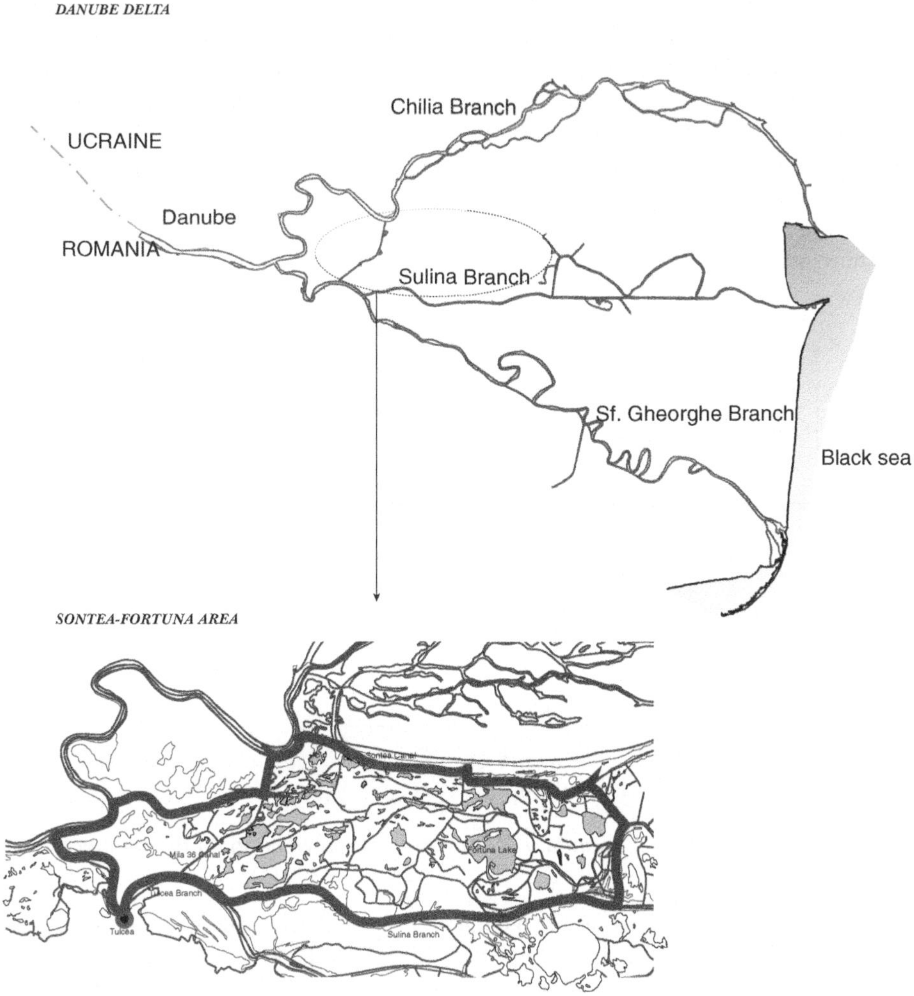

Figure 8.3 Sontea-Fortuna wetland system geographical location.

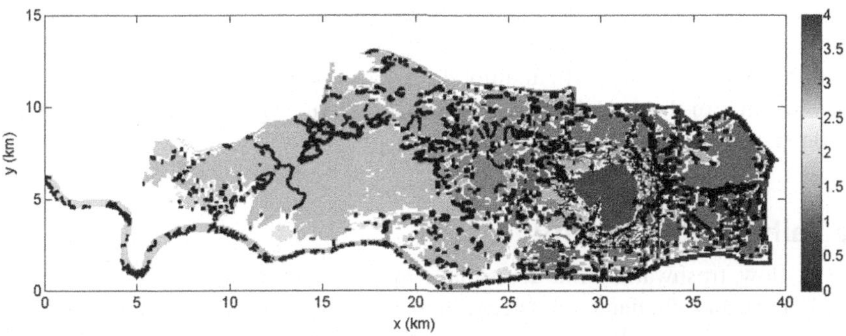

Figure 8.4 Water levels in Sontea Fortuna wetland for dry hydrological conditions.

Model is setup for three hydrological conditions; wet, normal and dry year, which correspond to high, medium, and low water level respectively, based on measured data kindly provided through the EU FP7 research project EnviroGRIDS (Popescu *et al.*, 2014). The Sontea–Fortuna area is modelled using Delft3D, which is a tool developed by Deltares (Deltares, 2010) to simulate hydraulic phenomena in river, estuarine and coastal areas. The software simulates variations in time and space (2D or 3D) of hydrodynamics, morphology water quality and sediment transport phenomena. The Sontea–Fortuna grid is a combination of grid size cells that varies from 90 m × 60 m to 160 m × 90 m. Total number of computational cells is 37039.

Figure 8.4 shows water levels in the area, as results of the model, for dry hydrological conditions.

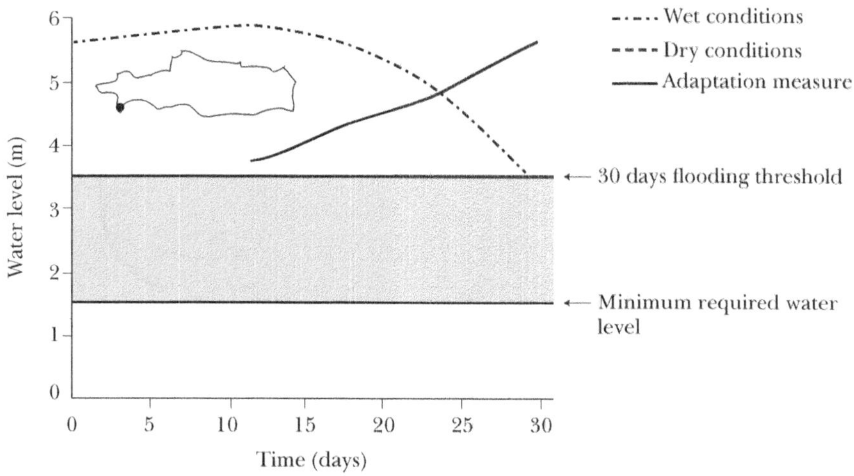

Figure 8.5 Water levels at upstream boundary of Sontea-Fortuna wetland for dry hydrological conditions.

Results help in identifying dry areas, where habitat is at risk and water quality conditions deteriorate, for too low water levels. Treshold for water levels are set by policies of the Romanian Ministry of Environment.

Figures 8.5, 8.6 and 8.7 shows results of water levels predictions in three points of the Sontea-Fortuna wetland for the three modelling cases; wet condition (October–December, 2010), dry condition (June–August, 2010) and one adaptation measure for the case of dry condition. The three points are located in the upstream, middle and downstream of the wetland, respectively.

For the three considered cases it can be seen that water level can be maintained between and above the levels of 1.5 m (m.b.s.l) to 3.5 m (m.b.s.l), and flooding does not exceed 30 days.

Based on the outcomes of the model adaptation measures can be proposed for improving the ecological status of the wetland.

8.5.2 Lake Taihu water quality

Lake Taihu is a shallow, freshwater wind-driven lake in the Yangtze Delta, located at the border between the provinces of Jiangsu and Zhejiang in Eastern China (see Figure 8.8). Eutrophication is always a serious problem in Taihu Lake. Water pollution associated with population increase, massive industrial growth with highly polluting heavy industry has taken a serious toll throughout China (Boqiang, 2007).

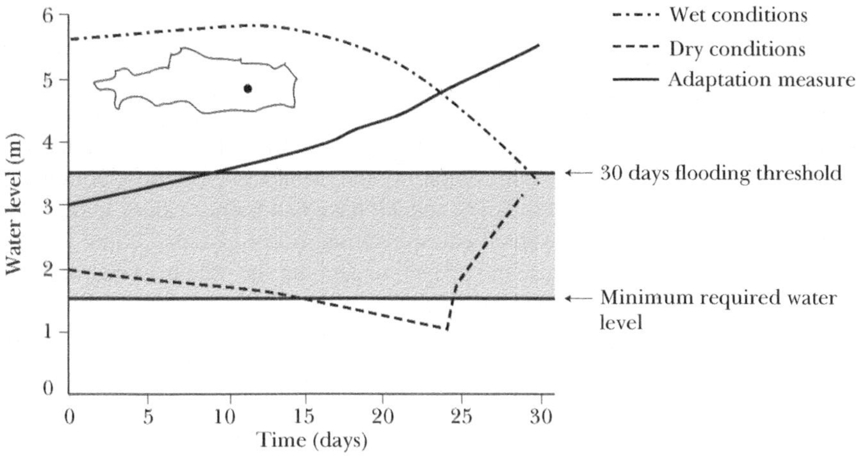

Figure 8.6 Water levels in the middle of Sontea-Fortuna wetland for dry hydrological conditions.

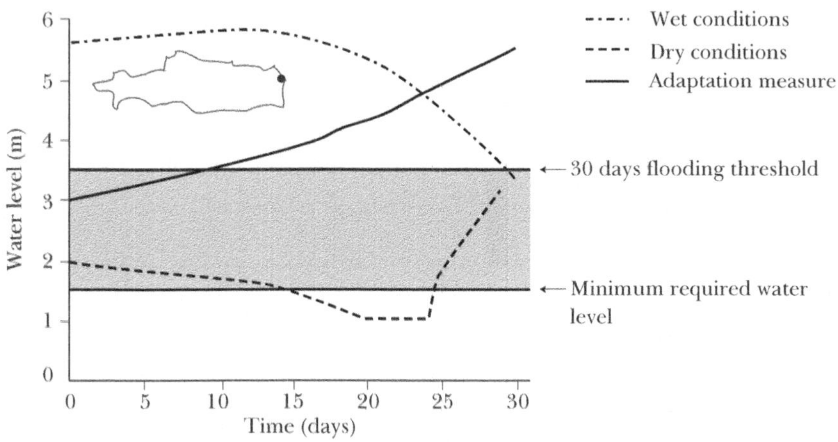

Figure 8.7 Water levels at downstream boundary of Sontea-Fortuna wetland for dry hydrological conditions.

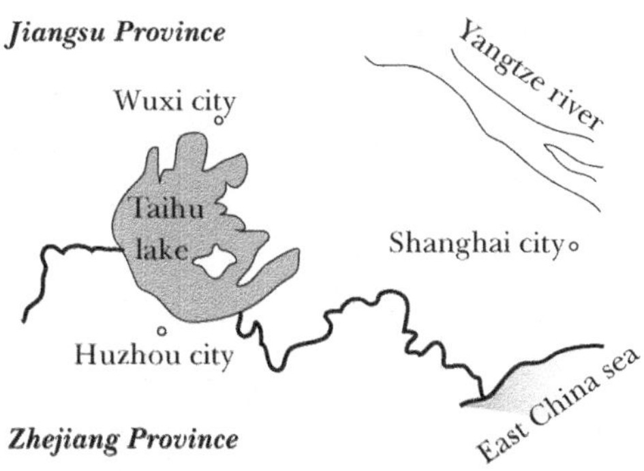

Figure 8.8 Lake Taihu location.

Many studies have been carried out regarding the water quality condition in the Taihu Basin. Zhu and Geng (2004) show that according with Chinese surface water quality standards there are 5.99 billion m^3 of water (42% in the total runoff) that are severely polluted and unusable for any user in Lake Taihu. Wang *et al.* (2011) estimated the pollutant fluxes of rivers in Zhejiang Province as part of the Taihu Lake Basin and indicated that the increasing levels of inflow of pollutants in the lake is the main reason for the deterioration of water quality in the lake. When comparing the mean values to those obtained between 1991 and 2001, the water quality parameters such as TN and TP have had higher values in the recent five years. Mao *et al.* (2008) published a three-dimensional eutrophication model of Lake Taihu, which fully couples the biological processes and hydrodynamics, and takes into account the effects of sediment release and the external loads from the tributaries. The main outcome of such a model is that two of the Lake Taihu's bays; Zhushan Bay and Xu Bay; are susceptible to algal blooms with high chlorophyll-a concentration.

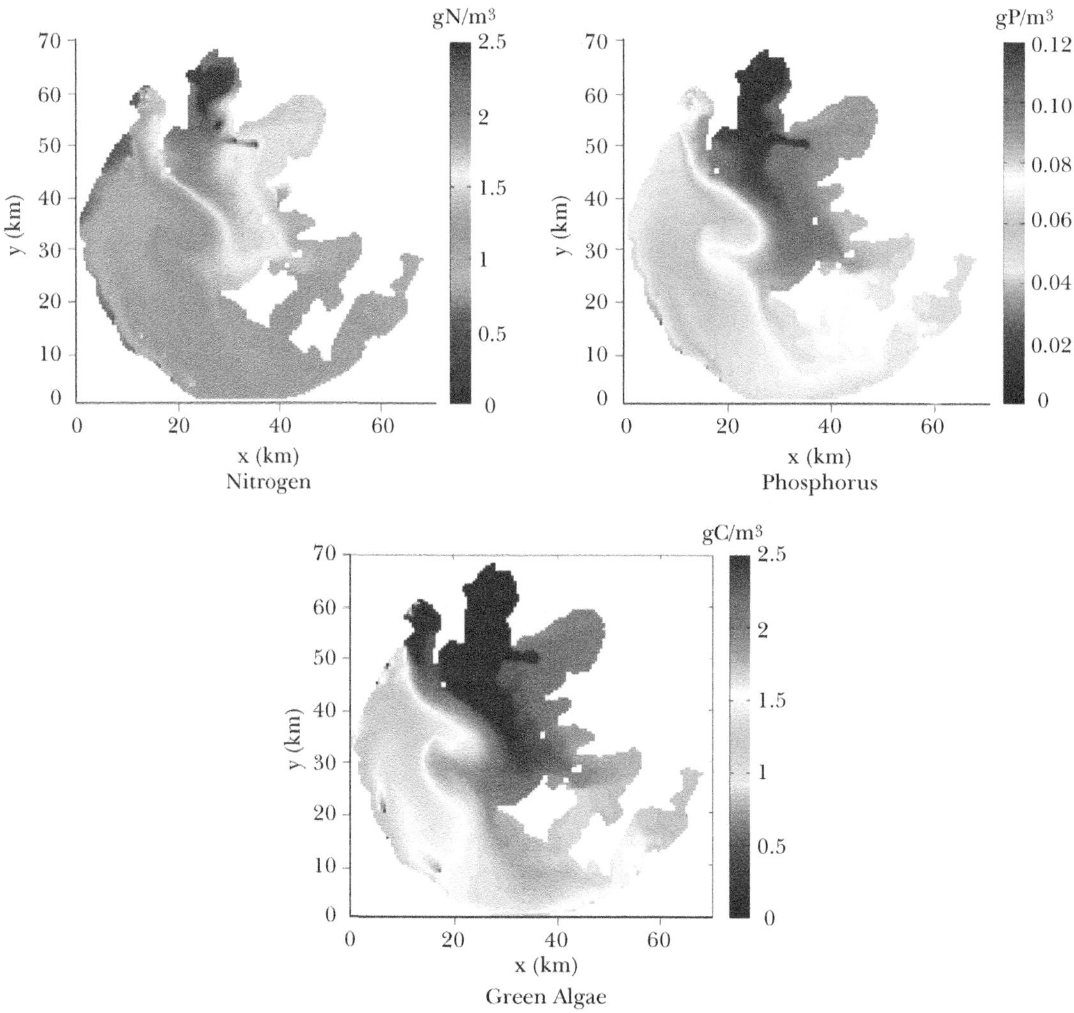

Figure 8.9 Spatial distribution of Nitrogen and Phosphorus in Taihu lake, during year 2008.

Here are presented water quality model results using a two-dimensional hydrodynamic model coupled with a two-dimensional water quality model. Model is set-up to investigate possible effects of wind on the eutrophication of the lake. A Delft3D model is built to simulate the hydrodynamic state of the lake. Hydrological, topographical and water quality data to set-up an example model are collected from published papers of Yue *et al.* (2013); Hu *et al.* (1998); Zhu and Geng (2004) and Yiping and Zhougho (2011). Three different sets of wind conditions (no wind, constant wind and measured wind) are considered as wind effects on the water level and velocity, in wet and dry seasons. Based on the calibrated hydrodynamic model a two dimensional water quality model is built and calibrated based on estimated data inputs for nutrient loads. A one-year hydrodynamic simulation is performed, for the year 2008, in order to show possibilities for determining the state of water quality in Lake Taihu. The estimated nutrient loads included total nitrogen (TN) and total phosphorus (TP) concentrations. The estimates are based on the population of cities around Lake Taihu, and the wastewater treatment plant capacity of the provinces of Jiangsu and Zhejiang, where the main sources of pollutants come from. Two time periods (spring and autumn) of algae blooms development for the year 2008 are noticed to form in the lake.

Figures 8.9 and 8.10 shows different modelling results for nitrogen, phosphorus and algae bloom for Taihu Lake, for the year 2008, for conditions of constant wind conditions.

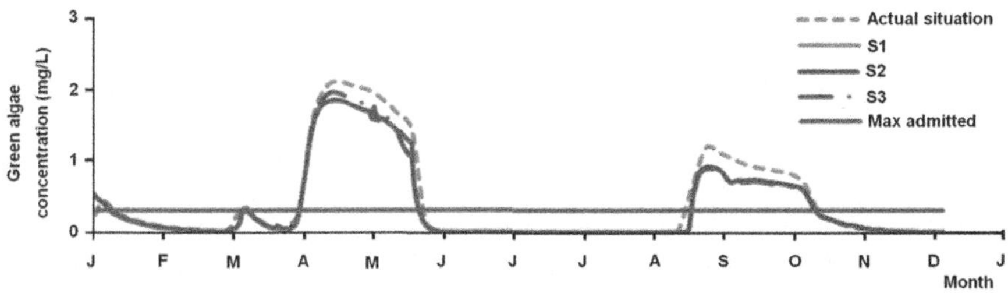

Figure 8.10 Temporal distribution of Algae bloom in Taihu lake, during year 2008.

The results of the study indicate that the hydrodynamic model with constant wind conditions (southeast, 5 m/s) shows better results than the one where measured wind conditions are used, simply because the measured wind data is insufficient; and an increase in population in the area would lead to an increase in the maximum value of algae concentration from 35% to 76%. An improvement of technologies used for the treatment of the wastewater may bring the maximum value of algae concentration between 17% and 42%.

References

Abbot M. B., Minns A. W. (1998). Computational Hydraulics, 2nd ed, Alder, Ashgate Publishing, Aldershot, Hampshire, UK.

Abbott M. B. (1979). Computational Hydraulics. Pitman, London.

Abbott M. B. and Basco D. R. (1989). Computational fluid dynamics. Longman Scientific and Technical, New York.

Acreman M. C., Riddington R. and Booker, D. J. (2003). Hydrological impacts of floodplain restoration: a case study of the River Cherwell, UK. *Hydrological and Earth System Sciences*, **7**(1), 75–85.

Aldama A. A. (1990). Filtering Techniques for Turbulent Flow Simulation, Springer Lecture Notes in English 56. Springer, Berlin.

Anderson B. G., Rutherfurd I. D. and Western A. W. (2006). An analysis of the influence of riparian vegetation on the propagation of flood waves. *Environmental Modelling and Software*, **21**, 1290–1296.

Ascher U. M. and Petzold L. R. (1998). Computer Methods for Ordinary Differential Equations and Differential Algebraic Equations. SIAM, Philadelphia, PA.

Barendregt A. and Bio A. M. F. (2003). Relevant variables to predict macrophyte communities in running waters. *Ecological Modelling*, **160**, 205–217.

Bathurst S. C. (1986). Sensivity analysis of the Systeme Hydrologique Europeen for an upland catchment. *Journal of Hydrology*, **87**, 103–123.

Bellinger G. (2004). An Introduction to Modeling & Simulation. http://www.systems-thinking.org/modsim/modsim.htm (accessed 31 December 2013).

Bergstrom S. (1991). Principles and confidence in hydrological modelling. *Nordic Hydrology*, **22**, 123–136.

Boqiang Q. P. X., Qinglong W., Liancong L. and Yunlin Z. (2007). Environmental issues of Lake Taihu, China. *Hydrobiologia*, **581**, 3–14.

Boubez T. I., Funnell W. R. J., Lowther D. A., Pinchuk A. R. and Silvester P. P. (1986). Mesh generation for computational analysis. *Computer-Aided Engineering Journal*, **3**(5), 196–201.

Boyd J. P. (1989). Chebyshev and Fourier Spectral Methods. Lecture Notes in Engineering. Springer-Verlag, New York.

Branson F. A., Gifford G. F., Renard K. G. and Hadley R. F. (1981). Rangeland Hydrology. Kendall/Hunt, Dubuque, USA, p. 339.

Bredenhoeft J. D. and Konikow L. F. (1983). Ground-water models: Validate or invalidate. *Ground Water*, **31**(2), 178–179.

Brostow W. and Dussault J. P. (1978). Construction of Voronoi polyhedral. *Journal of Computational Physics*, **29**, 81–92.

Brunner G. W. (2010). HEC-RAS, River Analysis System Hydraulic Reference Manual. U.S. Army Corps of Engineers Hydrologic Engineering Center.

Buell W. R. and Bush B. A. (1973). Mesh generation – a survey. *Trans. ASME, Journal of Engineering Industry Series B*, **95**(1), 332–338.

Burnett A. and Watson I. (1995). Hydrology, an Environmental Approach. Lewis Publishers, Florida, USA, pp. 585–602.

Butcher J. C. (1987). The Numerical Analysis of Ordinary Differential Equations. John Wiley and Sons Inc., NY.

Castro-Gama M. E., Popescu I., Li S., Mynett A. and van Dam A. (2014). Flood inference simulation using surrogate modelling for the Yellow River multiple reservoir system. *Environmental Modelling and Software*, **55**, 250–265.

Casulli V. (1990). Semi-implicit finite-difference methods for the two-dimensional shallow water equations. *Journal of Computer and Physics*, **86**, 56–74.

Chapra S. C. and Canale R. P. (2006). Numerical Methods for Engineers, McGraw-Hill, New York.

Chau K. W. (1990). Application of the Preissmann scheme on flood propagation in river systems in difficult terrain. In: Hydrology in Mountainous Regions. I – Hydrological Measurements; the Water Cycle, H. Lang and A. Musy (eds), IAHS Press, Wallingford, pp. 535–543. IAHS Publication No. 193. ISBN: 0-947571-57-4.

Chaudhry M. H. (1987). Applied Hydraulic Transients. Van Nostrand Reinhold Company, New York.

Chow W. T., Maidment D. R. and Mays L. W. (1988). Applied Hydrology. McGraw-Hill, New York, p. 572.

Chintu L. (1986). Numerical modeling of unsteady open-channel flow. *Advances in Hydroscience*, **14**, 161–333.

Clemmens A. J., Holly F. M. and Schuurmans W. (1993). Description and evaluation of program Duflow. *Journal of Irrigation and Drainage Engineering, ASCE*, **119**(4), 724–734.

Cunge J. A. (1969). On the subject of a flood propagation method. *Journal of Hydraulic Research*, **7**, 205–230.

Cunge J. A., Holly F. M. and Verwey A. (1986). Practical Aspects of Computational River Hydraulics. Pitman Publishing, London, p. 250.

Cox B. A. (2003). A review of currently available in-stream water-quality models and their applicability for simulating dissolved oxygen in lowland rivers. *Science of the Total Environment*, **314–316**(1), 335–377.

Dahlquist G. and Bjorck A. (2007). Numerical Methods in Scientific Computing. SIAM, Philadelphia, USA.

de Saint-Venant B. A. (1871). *Theorie du mouvement non-permanent des eaux avec application aux crues des rivieres et a l'introduction des marees dans leur lit* (Theory of unsteady water flow with application to river floods and to propagation of tides in river channels). *Comptes rendus, Paris Academy of Science,* **73**, 148–154, 237–240.

Deltares (2010). DELFT-WAQ User Manual, Delatres, p. 250.

DHI (2012). MIKE 11: A modelling system for rivers and channels, reference manual. DHI Water & Environment, p. 524.

Djordjevic S., Prodanovic D. and Walters G. A. (2004). Simulation of transcritical flow in pipe/channel networks. *Journal of Hydraulic Engineering*, **130**(12), 1167–1178.

Dooge J. C. I. (1986). Looking for hydrologic laws. *Water Resources Research*, **22**, 46–58.

Durran D. (1998). Numerical Methods for Wave Equations in Geophysical Fluid Dynamics. Springer, New York, USA.

Durran D. R. (1999). Numerical Methods for Wave Equations in Geophysical Fluid Dynamics. Springer, New York.

EU WFD, Water Framework Directive (2000). Directive 2000/60/EC of the European Parliament and of the Council, Establishing a Framework for Community Action in the Field of Water Policy. The European Parliament, Brussels, Belgium.

Eykhoff P. (1974). System Identification: Parameter and State Estimation. John Wiley & Sons, London.

Fletcher C. A. J. (1991). Computational Techniques for fluid Dynamics, 1: Fundamental and General Techniques. Springer, Berlin.

Forsius J. and Huttula T. (1982). Application of a mathematical model to a branched watercourse. *Geophysica*, **19**(1), 55–64.

Garabedian P. R. (1966). Partial Differential Equations. John Willey and Sons, NY, p. 672.

Gichamo T. Z., Popescu I., Jonoski A. and Solomatine D. (2012). River cross-section extraction from the ASTER global DEM for flood modeling. *Environmental Modelling and Software*, **31**, 37–46.

Godlewski E. and Raviart P.-A. (1996). Numerical approximation of hyperbolic systems of conservation laws. *Applied Mathematical Sciences,* **118**, Springer-Verlag, New York, USA.

Guinot V. (2003). Godunov-type Schemes: An Introduction for Engineers. Elsevier, Amsterdam, NL.

Heathcote I. W. (2009). Integrated Watershed Management: Principles and Practice, 2nd ed., John Wiley & Sons, Inc., NY

Henderson F. M. (1966). Open Channel Flow. Macmillan Series in Civil Engineering, Macmillan Company, New York.

Hicks N. S., Smith J. A., Miller A. J. and Nelson P. A. (2005). Catastrophic flooding from an orographic thunderstorm in the central Appalachians. *Water Resources Research*, **41**, 1–17.

Hirch C. (1988). Numerical Computation of Internal and External Flows, Volume 1: Fundamental of Numerical Discretization. Wiley and Son, Chichester.

Ho-Le K. (1988). Finite element mesh generation methods: A review and classification. *Computer-Aided Design*, **20**(1), 27–38.

Horrit M. S., Bates P. D. (2002). Evaluation of 1D and 2D numerical models for predicting river flood inundation. *Journal of Hydrology*, **268** , 87–99.

Horritt M. S. (2000). Calibration of a two-dimensional finite element flood flow model using satellite radar imagery. *Water Resources Research*, **36**(11), 3279–3291.

Holly F. M. Jr. and Preissmann A. (1977). Accurate calculation of transport in two dimensions. *Journal of Hydraulic Engineering*, **3**(11), 1259–1277.

Hu W. P., Pu P. M. and Qin B. Q. (1998). A three-dimensional numerical simulation on the dynamics in Lake Taihu, China (II): The typical wind-driven current and its divergence. *Journal of Lake Science*, **10**(4), 26–34.

Huges D. A. and Beater A. B. (1987). A comparison between lumped and semi-distributed approaches to modelling isolated flood events. In: Proceedings of the Hydrological Sciences Symposium, D. A. Hughes and A. W. Stone, (eds), Departement of Geography, Rhodes University, Grahamstown, RSA, pp. 547–561.

Hunsaker C. T. and Levine D. A. (1995). Hierarchical approaches to the study of water quality in rivers. *BioScience*, **45**,193–203.

ILEC (2005). Managing Lakes and their Basins for Sustainable Use: A Report for Lake Basin Managers and Stakeholders. International Lake Environmental Committee Foundation, Kusatsu, Japan.

Imberger J. (1987). Mixing processes relevant to phytoplanktondynamics in lakes. New Zealand *J. Marine Freshwater Res.* **21**, 361–377.

Jain S. C. (2001). Open-Channel Flow. John Wiley and Sons, New York.

James L. D. and Burges J. J. (1982). Selection, calibration and testing of hydrologic medels. In: Hydrologic Modeling of Small Watersheds, C. T. Haan, H. P. Johnson and D. L. Brakensiek (eds), ASAE Monograph No. 5, St. Joseph, Michgan, pp. 437–472.

Jung N.-C., Popescu I., Kelderman P., Solomatine D. P. and Price R. K. (2010). Application of model trees and other machine learning techniques for algal growth prediction in Yongdam reservoir, Republic of Korea. *Journal of Hydroinformatics*, **12**(3), 262–274.

Jung N.-C., Popescu I., Price R. K., Solomatine D., Kelderman P. and Shin J.-K. (2011). The use of the A.G.P. test for determining the phytoplankton production and distribution in the thermally stratified reservoirs: The case of Yongdam reservoir in Korea. *Environmental Engineering and Management Journal*, **10**(11), pp. 1647–1657.

Knight D. W. (2006). Introduction to flooding and river basin modelling, Chapter 1. In: River Basin Modelling for Flood Risk Mitigation D. W. Knight and A. Y. Shamseldin (eds), Taylor & Francis, 1–19.

Lazarov R. D., Makarov V. L. and Samarskii A. A. (1982). Application of exact difference schemes to the construction and study of difference schemes for generalized solutions. *Matematisk Sb. (N.S.)*, **117**(4), 469–480.

Lane L. J., Nichols M. H., Hernandez M., Manetsch C. and Osterkamp W. R. (1994). Variability in discharge, stream power, and particle-size distribution in ephemeral-stream channe systems, in Variability in Stream Erosion and Sediment Transport (Proc. of the Canberra Symposium December 1994). Edited by Olive, L. J. Loughram, R. J. and Kesby, J. A. IAHS Publ. no. 224, 335–342.

Legras J. (1971). *Méthodes et Techniques de L'analyse Numérique* (Methods and Techniques for Numerical Analysis). Dunod, Paris.

Lesieur M. (1997). Turbulence in Fluids, Fluid Mechanics and its Applications, 40, 3rd edn, Kluwer Academic Publishers, Dordrecht, The Netherlands.

LeVeque R. J. (2002). Finite Volume Methods for Hyperbolic Problems (Cambridge Texts in Applied Mathematics). Cambridge University Press, Cambridge, UK.

LeVeque R. J. (2007). Finite difference methods for ordinary and partial differential equations: Steady-state and time dependent problems, SIAM, Philadelphia, PA.

Liggett J. A. and Cunge J. A. (1975). Numerical methods of solution of the unsteady flow equations. In: Unsteady Flow in Open Channels, K. Mahmood and V. Yevjevich (eds), Water Resources, Fort Collins, Colorado, pp. 89–182.

Louks, D. P. and van Beek E. (2005). Water resources systems planning and management. Studies and Reports in Hydrology, pp. 649–672.

Mandel J., McCormick S. C. and Multilevel A. (1989). Variational Method for Au = λBu on Composite Grids. *Journal of Computational Physics*, **80**, 442–452.

Manteuffel T. A. and White A. B. Jr. (1987). The numerical solution of second-order boundary value problems on nonuniform meshes. *Mathematics of Computation*, **47**, 511–535.

Mao J., Chen Q. and Chen Y. (2008). Three dimensional eutrophication model and application to Taihu Lake, China. *Journal of Environmental Sciences*, **20**, 278–284.

Martin J. L. and McCutcheon S. C. (1999). Hydrodynamics and Transport for Water Quality Modeling. CRC Press, Inc, 335–384.

Meselhe E. A. Holly F. M. Jr. (1997). Invalidity of Preissmann scheme for transcritical flow. *Journal of Hydraulic Engineering*, **123**(7), 652–655.

Moya Quiroga V., Popescu I., Solomatine D. P. and Bociort L. (2013). Cloud and cluster computing in uncertainty analysis of integrated flood models. *Journal of Hydroinformatics*, **15**(1), 55–70.

Novak P., Moffat A. I. B., Nalluri C. and Narayanan R. (1990). Propagation of translatory waves in channels and rivers 4th edn, Editeed by the Academic Division of Unwin Hyman Ltd.

Ogden F. L. and Julien, P. Y. (2002). Mathematical Models of Small Watershed Hydrology and Applications. 1st ed. Colorado, US, Water Resources Publication, LLC, 90–91. (eBook).

Orlob G. T. (1983). Mathematical Modeling of Water Quality Streams, Lakes, and Reservoirs. Wiley and Sons, New York.

Ponce V. M. and Simon D. B. (1997). Shallow water propagation in open channel flow. *Journal of Hydraulic Division, ASCE*, **103**, 1461–1476.

Popescu I. (2013). Unsteady Flow in Floods in a changing climate: Inundation Modelling G. di Baldassarre, Cambridge University Press, Cambridge, UK.

Popescu I., Jonoski A. and Bociort L. (2012). Decision support systems for flood management in the Timis Bega catchment. *Environmental Engineering and Management Journal*, **11**(12), 2305–2311.

Preissmann A. (1961). Propagation of translatory waves in channels and rivers., Proc., 1st Congress of French Association for Computation, Grenoble, France, A. F. C. A. L., 433–442 (in French).

Price R. K. (1985). Flood routing. Chap 4 in Developments in Hydraulic Engineering-3. Elsevier Applied Sciences Publishers, London, pp. 129–174.

Rauch W., Henze M., Koncsos L., Reichert P., Shanahan P., Somlyody L. and Vanrolleghem P. (1998). River water quality modelling: I. State of the art. *Water Science and Technology* **38**(11), 237–244.

Richtmyer R. D. and Morton K. W. (1967). Difference methods for initial-value problems. Interscience Publishers, New York.

Rouse H. (1950). Fundamental Principles of Flow. In: Engineering Hydraulics. H. Rouse (ed.), Wiley, New York, USA, p. 135.

Sagaut B. (2001). Large Eddy Simulation for Incompressible Flows. Springer-Verlag, Berlin Heidelberg, New York.

Schaffer H. A., Madsen P. A. and Deigaard R. (1993). A Boussinesq model for waves breaking in shallow water. *Coastal Engineering*, **20**, 185–202.

Schultze R. E. (1995). Hydrology and Agrohydrology. Water Research Commision, Pretoria, RSA, WRC Report TT69/95, p. 552 (ISBN: 1-86845-136-4).

Shepherd R. G. and Geter W. F. (1995). Verification, calibration, simulation: Protocolls in ground-water and AG/NP5 modeling. In: Water Quality Modeling, C. Heatwole (ed.), American Society of Agricultural Engineers, St. Joseph, MI, USA, pp. 87–90.

Singh V. P. (ed.) (1995). Computer Models of Watershed Hydrology. Water Resources Publication, Highlands Ranch CO, USA, p. 1130.

Singh V. P. (1996). Kinematic wave modeling in water resources, Surface Water Hydrology, John Wiley & Sons Inc., New York

Smith M. and Kohn R. (2002). Parsimonious covariance matrix estimation for longitudinal data. *J. Amer. Statist. Assoc.*, **97**, 1141—1153.

Sorooshian S. (1991). Parameter estimation, model identification, and model validation: Conceptual-type models. In: Recent Advances in the Modeling of Hydrologic Systems, D. S. Bowles and P. E. O'Connell (eds), Kluwer Academic Publishers, Dordrecht, Netherlands, pp. 443–467.

Strelkoff T. and Katodes N. D. (1977). Dimensionless solutions of border-irrigation advance. *Journal of Irrigation and Drainage Division, ASCE*, **103**(4), 401–417.

Streeter H. W. and Phelps E. B. (1925). A Study of the pollution and natural purification of the Ohio river. III. Factors concerned in the phenomena of oxidation and reaeration, Public Health Bulletin no. 146, Reprinted by U.S. Department of Health, Education and Welfare, Public Health Service, 1958.

Sturm T. W. (2001). Open Channel Hydraulics. McGraw-Hill Series in Water Resources and Environmental Engineering, New York.

Szymkiewicz R. (1993). Solution of the advection-diffusion equation using the spline function and finite elements. *Communication Numeral Methods Engineering*, **9**(4), 197–206.

Szymkiewicz R. (1995). Method to solve 1D unsteady transport and flow equations. *Journal of Hydralic Engineering*, **121**(5), 396–403.

Szymkiewicz R. (2010). Numerical Modeling in Open Channel Hydraulics. Springer, Dordrecht, p. 370.

Thacker W. C. (1980). A brief review of techniques for generating irregular computational grids. *International Journal of Numerial Method Engineering*, **15**, 1335–1341.

Tim U. S. (1995). Emerging issues in hydrologic and water quality modeling research. In: Water Quality Modeling, C. Heatwolw, (ed.), American Society of Agricultural Emgineers, St. Joseph, MI, USA, pp. 358–373.

Todini E. and Bossi A. (1986). Parabolic and Backwater, an unconditionally stable flood routing scheme particularly suited for real time forecasting and control. *Journal of Hydraulic Research*, **24**(5), 405–424.

van Leer B. (1979). Towards the Ultimate Conservative Difference Scheme V. A second order sequel to Godunov's method. *Journal Computational Physics*, **32**(1), 101–136.

Wang F. E., Tian P., Yu J., Lao G. M. and Shi T. C. (2011). Variations in pollutant fluxes of rivers surrounding Taihu Lake in Zhejiang Province in 2008, *Physics and Chemistry of the Earth*, **36**, 366–371.

Wetzel R. G. (2001). Limnology: Lake and Reservoir Ecosystems. Academic Press, New York.

Wright N. G. and Crosato A. (2011). The hydrodynamics and morphodynamics of rivers, in Treatise on Water Science, Four-Volume Set P. Wilderer, Editor. Elsevier Science Ltd. pp. 135–156.

Woods R. A., Sivapalan M. and Duncan M. (1995). Investigating the representative elementary area concept: An approach based on field data. In: Advances in Hydrological Processes, Scale Issues in Hydrological Modelling, J. D. Kalma and M. Sivapalan (eds), John Wiley and Sons, UK, pp. 49–70.

Woolhiser D. A. and Brakensiek D. L. (1982). Hydrologic modeling of small watersheds. In: Hydrologic Modeling of Small Watersheds, C. T. Haan, H. P. Johson and D. L. Brakensiek (eds), ASAE Monograph 5, St. Josephs, MI, USA, pp. 3–16.

Wu W. (2008). Computational River Dynamics. Taylor & Francis, London, p. 494, ISBN: 978-0-414-44960-1.

Yiping L. K. A. and Zhongbo Y. (2011). Modeling impacts of Yangtze River water transfer on water ages in Lake Taihu, China. *Ecological Engineering*, **37**, 325–334.

Yue C., Popescu I., Mynett A., Pan Q. and Postma L. (2013). Challenges for 2D water quality modelling of lake Taihu in China. *Environmental Engineering and Management Journal*, **12**(5), 1031–1044.

Zagona E. A., Fulp T. J., Shane R., Magee T. M. and Goranflo H. M. (2001). RiverWare: A generalized tool for complex reservoir system modeling. *Journal of the American Water Resources Association*, **37**(4), 913–929.

Zalewski M. and Wagner-Lotkowska I. (eds). (2004). Integrated Watershed Management – Ecohydrology and Phytotechnology-Manual. UNESCO IHP, UNEP IETC. pp. 246.

Zhu W. and Geng Y. Q. (2004). Water quality issues in the catchment water balance of Taihu Lake, China. In: Hydrology: Science and Practice for the 21st century, Volume II, Proceedings of the British Hydrological Society International Conference, Imperial College, London, July 2004, 534–541.

Index

IWA Publishing's authorised EU representative for General Product Safety Regulations is Diane D'Arras, 15 rue Duret, 75116 Paris, France, e-mail: safety@iwap.co.uk.

Printed and bound by CPI Group (UK) Ltd, Croydon, CR0 4YY
12/05/2026
02108891-0018